Smallpox : Cholera

VASILE TUDOR, M.D. : **IOAN STRATI**, M.D.

ABACUS PRESS

Tunbridge Wells, KENT

1977

SMALLPOX : CHOLERA

ABACUS PRESS
Abacus House, Speldhurst Road, Tunbridge Wells
Kent, England

ISBN 0 85626 092 4

Authorised, revised, updated translation from the Romanian language edition of "VARIOLA. HOLERA" first published by **EDITURA MILITARĂ**, 1971.

Editor: Colonel ing. DORIN DIMITRIU
Translated by **IOANA STURZA**
Translation edited by **S.W. BOOTHE** and **JOHN HAMMEL** Ph. D.

Printed in Romania

Contents

Part II

CHOLERA

PREFACE TO THE ENGLISH EDITION

The publication of this English translation of our work has given us the opportunity of carrying out a thorough revision and updating of the material presented in the first, the Romanian, edition. English readers thus have, at the time this goes to press, the latest available data concerning smallpox and cholera.

V. TUDOR : I. STRATI

INTRODUCTION

This monograph is the result of experience culled in the course of a study tour in India and is designed to present to interested readers a general survey of the present-day problems posed by outbreaks of smallpox and cholera. Some aspects of the problems have seen few changes in the course of the years, but others such as epidemiology, clinical work and the treatment of cholera are at present undergoing a basic review.

During our stay in India we studied at various institutions and research centres where we were able to observe the methods used by specialist staff for diagnosing smallpox and cholera, and the stratagems employed in the biological control of these infections. The following are some of the institutes visited: All India Institute of Medical Sciences (New Delhi), National Institute of Communicable Diseases (New Delhi), Cholera Research Centre (Calcutta), Indian Institute of Experimental Medicine (Calcutta), School of Tropical Diseases (Calcutta), Central Research Institute (Kasauli), Haffkine Institute (Bombay), Virus Research Centre (Poona), King Institute (Madras).

We were able, also, to study in several hospitals (the Hospital of Communicable Diseases of Calcutta; the Kasturba Hospital of Communicable Diseases of Bombay; the Tondiarped Hospital of Madras) and gained first-hand experience at the bedsides of some of the large number of patients that can be met with in India, where these two diseases, in spite of all the preventive measures that are taken, evolve endemo-epidemically. We saw, too, the work of the anti-epidemic services: on the one hand, the central services' organisation and methodology for combating cholera and, on the other, the toilers in the field who daily are in close contact with the disease.

This book, written by a clinician and a laboratory specialist, has been conceived for the use of public health officers responsible for problems of organisation, particularly those concerned with infectious diseases: epidemiologists, clinicians, laboratory workers. It is intended not only to provide a background knowledge of these diseases, of which little practical experience has been possible to physicians in the western hemisphere, but also detailed organisational and clinical information that will prove invaluable should either or both these diseases make an unwanted appearance in the West.

We wish to express our gratitude to our colleagues in India who, with the kindness and generosity characteristic of the Indian people, passed on to us unstintingly so much of the knowledge they have gained in their lifelong struggle against the ravages of smallpox and cholera. In particular we would like to mention: Dr Shrivastav (New Delhi), Dr Mahendra Singh (New Delhi), Dr Raghavan (New Delhi), Dr Jhala (Bombay), Dr T. R. Rao (Poona), Dr Thomas (Kasauli), Dr Rao (Madras), Dr Desphande (Bombay), Dr Nimbkar (Bombay), Dr Karbhari (Bombay), Dr Mancharamani (Bombay), Dr Sarkar (Calcutta), Dr Roy (Calcutta), Dr Kothari (New Delhi) and Dr Gracias (Poona).

V. TUDOR: I. STRATI

PART I

Smallpox

1. General

Smallpox is an acute infectious disease caused by the variola virus. It is of exceptional gravity, extremely contagious, manifested clinically by a state of accentuated toxaemia — in the course of which fever is never absent — and a generalised vesiculo-pustulous eruption that after recovery leaves definitive marks.

According to the clinical aspect there are two forms of variola — VARIOLA MAJOR and VARIOLA MINOR (alastrim) — each with a geographical distribution of its own, which correspond to variola viruses that can be differentiated by laboratory tests. Both forms of variola are included in international health legislation.

Variola minor (alastrim) is encountered in South America and variola major in the rest of the world.

Alastrim has the same clinical and epidemiological evolution as the major form of smallpox but the symptoms are more attenuated and the mortality rate is very low.

In India the major form alone evolves, and the authentic clinical picture of the disease, as described many years ago, is encountered as a rule even when milder forms are recorded, as a consequence of the increased resistance of the population due especially to vaccination.

India represents one of the main endemic focuses of smallpox from which the disease has often spread to other countries; the number of cases registered in India represents approximately one third of all the cases throughout the world.

1. 1. History

Smallpox is a scourge that has affected mankind since ancient times, being known in India and China for approximately 1700 years before our era. One of the first descriptions of smallpox, then called Tai-Tou, comes from China and dates from 1122 B.C.

Epidemics have been described in 200 B.C. and 312 A.D. and the historians tell us that Mecca was saved in the year 569 when the Abyssinians who attacked it were mown down by smallpox.

Documents mention an epidemic outbreak of smallpox in the year 570 in France and Italy when the term variola was used for the first time; it derives either from *varius* (spotted) or *varus* (small button).

The Crusades favoured the spread of smallpox throughout Europe, which was swept by epidemics during the 15th and 16th centuries.

In the Americas smallpox appeared soon after Columbus's first voyage, and rapidly spread throughout Central and South America, killing thousands of the native population in 1520. The trade in slaves from Africa contributed to the prevalence of smallpox in North America.

Starting in the 17th century, the epidemic potential of smallpox significantly increased and the disease caused overwhelming desolation; it is considered that in this century the number of Europeans who died of smallpox rose to about 60 million. The number of victims was so great that an adage of those times said that *smallpox can only be avoided by he who does not live long enough to wait for it.*

The natural dread of the disease that filled the people stimulated various attempts to prevent it, based upon the empirical observation that those who pass through a mild form of smallpox do not contract the disease again. For this purpose, long before our era, the custom existed in China of putting on healthy children shirts smeared with the pus of smallpox patients or introducing in the nose old crusts ground into powder. In other regions coins smeared with the patient's pus were put under the children's axilla. In Armenia, Bengal and Abyssinia the smallpox crusts were swallowed, and in India and the Caucasus were inoculated with pus to healthy people by means of a handful of needles. In Russia, steam baths were taken, after which the people were beaten with brooms smeared with pus.

In 1718, the wife of the English ambassador to Turkey, Lady Mary Wortley Montague introduced in England the variolation procedure, which subsequently spread to other European countries, to France, the Romanian Principalities, Russia, etc. It consists of inoculation of old crusts by incision or puncture of the skin. It gives immunity but is very risky, often causing a severe disease with a mortality rate of more than four deaths per thousand "vaccinees"; for instance, in England, during the first eight years of variolation 17 of 897 persons inoculated died, and the practice was banned in 1840 by Parliament.

An important moment in the history of smallpox was the vaccination performed by Edward Jenner. Starting from the observation that people who have suffered from cowpox become immune to smallpox Jenner variolised 16 persons who had already suffered from cowpox and saw that they did not contract smallpox. On May 14, 1776, Jenner inoculated a child, James Philips, on the arm with the contents of a vaccinia pustule that had appeared on a cow-hand; after several days the child developed a characteristic vaccine pustule. One year later he inoculated the child on both arms with pus from a human smallpox lesion: the result was negative; a second inoculation gave similar results. In 1798 Jenner carried out new experiments on humans. He inoculated a child with products collected directly from cows and from this child performed a series of inoculations in other people. With the product obtained after five human passages Jenner inoculated several persons who resisted the subsequent administration of smallpox products.

Jenner published his results in the same year, his method consisting of the inoculation of vaccine from arm to arm, the content of the pustules on day 6 and 7 after inoculation being considered the most effective.

Jenner's method rapidly spread throughout Europe.

As collection of the vaccinal lymph from humans was limited, and inoculations from arm to arm transmitted other diseases (e.g. syphilis, tuberculosis, erysipelas) and also regenerated the virus maintained in man, it was necessary to resort once again to cattle, inoculating calves with human vaccine products.

This modified technique marked an important step in the prevention of smallpox since it insured sufficient amounts of vaccine and avoided the risks involved in vaccination from man to man.

The introduction of glycerine to conserve the vaccine lymph (in 1868) greatly contributed to the preparation of the modern antismallpox vaccine. Although marked improvements and changes have been brought to the preparation of the vaccine, actually today immunisation against smallpox is based upon the results obtained during the past century.

The introduction and spread of specific prevention has radically changed the epidemiology of smallpox. During the 19th century the incidence of smallpox significantly fell in Europe; there was a single severe epidemic in 1870—1871 helped by the Franco-Prussian war. The French army, insufficiently immunised had 280 000 cases of smallpox with 23 470 deaths, whereas among the German troops who had been vaccinated and revaccinated there were only 8360 cases with 297 deaths. In order to evaluate these figures it is sufficient to recall that about 500 000 persons died in India from smallpox during the 19th century.

Towards the end of the 19th century in Europe deaths from smallpox were exceeded only by those from scarlet fever and measles.

The 20th century marks the constant restriction of the area of smallpox. During the first half of the century smallpox was eradicated in Europe and North America. Following the application of a programme started in 1950 smallpox was likewise eradicated in Central and South America, except Brazil and Columbia.

During this period smallpox had disappeared from many countries in north Africa, Asia and the Near East. In 1958 the XIth World Health Assembly proposed to initiate a world wide programme for the eradication of smallpox, and in 1966 the XIXth World Health Assembly proposed an intensified programme. Following this decision many countries adopted a programme of systematic survey and vaccination of the population. In some countries no more variola cases were recorded, either permanently or at least for certain periods. The implementation of a close survey led to the notification of a greater number of cases, so that the general incidence of the disease appeared to be greater. In some countries, for example India and Pakistan, however, after a first period in which the number of cases fell, a marked recrudescence of smallpox occurred in 1967. Subsequently, after the control programme was ameliorated the incidence again fell.

1. 2. The present situation of smallpox in the world

Starting in 1967, with the application of the intensive programme proposed by the WHO for the eradication of smallpox there was a progressive, constant decrease in the number of cases (the number of countries that notified the presence of the disease within their territory and the morbidity rate in the endemic countries).

TABLE 1. 1.

Number of smallpox cases throughout the world from 1961 to 1974

Year	Number of cases
1961	88 761
1962	98 638
1963	132 773
1964	75 201
1965	112 155
1966	92 402
1967	131 418
1968	80 209
1969	54 223
1970	33 304
1972	65 000
1973	135 233
1974	163 634*

* Communicated to the WHO up to 9 July 1974.

As may be seen from Table 1.1, the number of cases decreased after 1967 but since 1971 the incidence of smallpox has shown a steady tendency to increase: in the first half of 1974 the number of cases notified to the WHO reached that of the whole of 1973.

In order to underline the difficulties encountered in the implementation of the smallpox eradication programme it should be recalled that in 1970 the data recorded led some specialists to foresee that in 1971 there would not be more than 21 000 variola cases in the world and that in no country (except perhaps Sudan) would the morbidity rate be higher than 5 per 100 000 inhabitants.

At any rate the results of the eradication programme are evident: the number of countries in which variola cases have been recorded has steadily decreased (61 in 1962, 43 in 1967 and 16 in 1971), as well as those in which the morbidity rate is higher than five per 100 000 inhabitants (15 in 1967 and five in 1970); smallpox was endemic in 30 countries in 1967 and in four countries in 1974.

The maps, Figs 1.1 and 1.2, give a clear picture of the reductions in smallpox cases that have been achieved since the WHO eradication programme was introduced.

●

As regards the evolution of smallpox there are two distinct categories: countries in which variola continues to be endemic and others

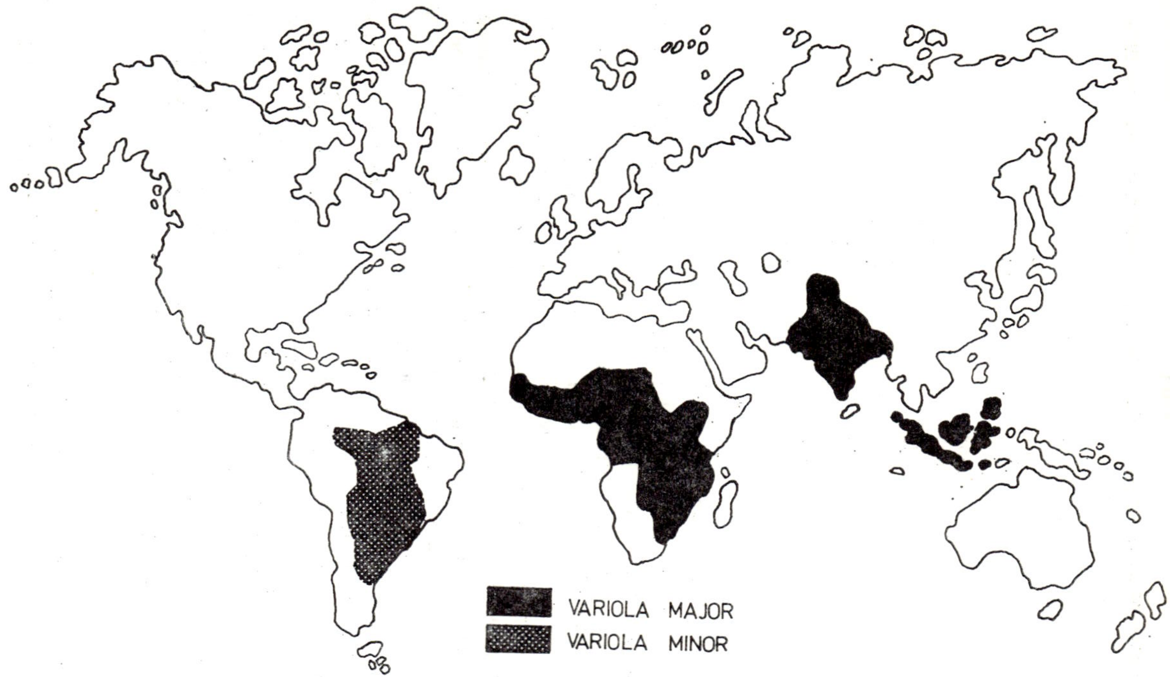

The spread of smallpox throughout the world (1970)
FIG. 1.1.

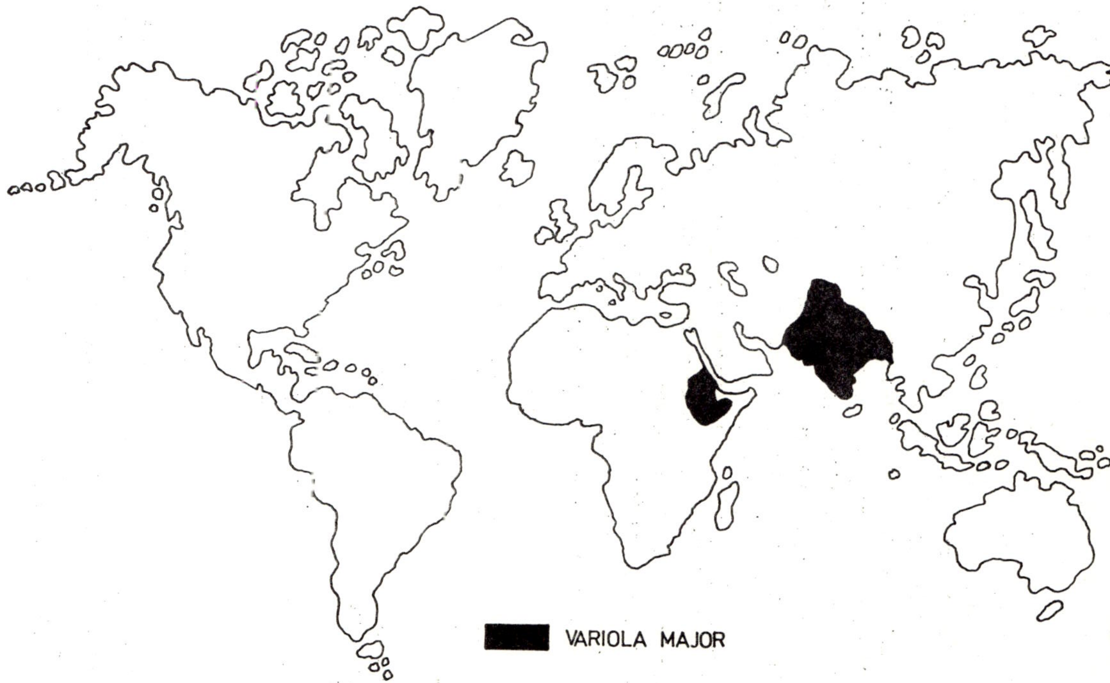

The spread of smallpox throughout the world (1975)
FIG. 1.2.

in which variola only appears in restricted foci initiated by cases imported from the former countries.

During the first six months of 1974 variola was still endemic in four countries: Bangladesh, India and Pakistan in Asia and Ethiopia in Africa. Bangladesh was apparently intact in August 1970 yet in 1972 there was a severe outbreak due to migration of the population, among the refugees returning from India. In 1972 there were more than 10 000 cases, in 1973 32 000 cases but at present the epidemic is regressing. In India morbidity from variola seems to have been increasing since 1971 (16 184 in 1971; 27 407 in 1972; 87 489 in 1973) and the forecasts are that this will continue in 1974. This increase in the number of cases is partly due to the improved survey but also to a recrudescence of the disease, especially in Bihar, Uttar Pradesh and Assam, with a potential danger of spread to the neighbouring regions. In Pakistan, too, there is a progressive increase in the number of cases (from 3192 in 1970 to 9258 in 1973).

In Ethiopia the incidence of variola continues to fall after reaching a peak in 1971 following improvement of the notification system.

Of greater importance, however, even in these countries with a high endemicity, is the fact that the affected areas are more and more restricted and transmission of the disease to other areas within these countries appears to have been practically arrested.

TABLE 1. 2.

Countries with endemic smallpox in 1970

Asia	Africa	America
India Indonesia Afghanistan Nepal Pakistan Bangladesh	Burundi Ethiopia Zaire Ruanda South Africa Rhodesia Sudan Tanzania Nigeria	Brazil

As compared to the situation in 1970 (see Table 1.2), during the first half of 1974 in at least two of the countries with endemic smallpox, transmission of the disease has been stopped: Brazil and Indonesia reported a 28 month period without any case of smallpox (the former in August 1973, the latter in April 1973).

Towards the end of 1973 the only countries that still notified a reduced number of cases were Afghanistan, Nepal, Botswana, Somaliland and the French territories of the Afars and Issas. In the first half of 1974 Nepal alone showed an increase in the incidence of smallpox cases (1051). For the countries of continental Asia and Africa one must still wait for the cases to be sufficiently spaced to be able to declare the full success of the smallpox eradication programme.

This brief review of the situation of endemic smallpox throughout the world shows that:

(1) From the distribution of smallpox cases reported to the WHO at the end of 1973 and from the morbidity curve in the first six months of 1974 it results that the main focus still is India (62% and respectively 85% of the total number of cases).

(2) The present foci in Nepal and Pakistan continue to evolve but to a much lesser extent.

(3) The foci in Bangladesh and in Ethiopia appear to be regressing.

(4) More than 90% of the cases notified throughout the world are to be found within a restricted area representing less than 15% of the surface of the four large endemic areas (Bangladesh, India, Pakistan, Ethiopia), underlining the success of survey and control and the adequacy of the vaccination campaign.

(5) Particular attention must be paid to the possible spread of the disease to the neighbouring districts, either within the country (in India from Bihar to the neighbouring districts) or to other countries (from India to Nepal and Bangladesh).

(6) It appears increasingly evident that eradication of variola is possible as there are more and more countries in which its transmission has been arrested. Although the predictions concerning the evolution of smallpox do not always come true, restriction of the epidemic towards the end of 1973 to a few areas may be considered very encouraging.

(7) It should not be overlooked that a smallpox epidemic may break out again in a former epidemic area (for instance Bangladesh).

(8) In the control of smallpox, survey of the areas affected seems to be the most important factor both for eradication and in the case of a new outbreak. In some countries the disease was eradicated before the end of the vaccination campaign and in others, in which survey activity had not yet started, transmission continues in spite of mass vaccination, regularly performed.

(9) The importance of vaccination of exposed groups results from the data which show that 80% of the smallpox cases appear in non-vaccinated persons and more than 80% in children under the age of 15.

Persistence of endemic smallpox within certain areas represents a permanent risk of contamination for the intact countries or those in which the eradication campaign was successful. Imported cases

TABLE 1. 3.

Smallpox cases in Europe between 1961 and 1970

Year	Country affected	Country from which smallpox was imported	No. of cases
1961	Spain	India	17
	West Germany	India	4
	U.S.S.R.	India	1
	Belgium	Congo	1
	West Germany	Liberia	5
	West Germany	Pakistan	33
	England	Pakistan	3
	England	Pakistan	2
	England	Pakistan	14
1962	England	Pakistan	1
	England	Pakistan	47
	Poland	India	29
	England	India	3
1963	Sweden	Asia	27
	Poland	India	99
	Switzerland	Gabon	1
	Hungary	—	2
1965	West Germany	Tanzania	1
1966	England	—	72
1967	West Germany	India	2
	Czechoslovakia	India	1
	England	Pakistan	2
1968	England	Pakistan	1
	Belgium	Congo	1
1970	West Germany	Pakistan	20
	Denmark	Afghanistan	1
	Norway	Denmark	1

and the outbreaks they generate is a frequently encountered cause of spread of the disease.

Between 1948 and 1968 there were 80 imported cases in intact countries. In Europe between 1961 and 1970 (Table 1.3) there were 28 imported cases which caused a total of 391 cases.

As may be seen from Table 1.3 most cases (20 of 28 cases) introduced into Europe came from Asia, especially India and Pakistan; five cases came from Africa, two from other European countries and in two cases the origin of the infection could not be established (1963 Hungary and 1966 England).

Most of the cases were imported by people travelling by air (23 out of 26 cases). The journey by sea generally takes longer and the disease develops before arriving in port; 25 of 29 such patients belonged to the crew. On the other hand, patients in the incubation stage may travel by air, and the traveller may leave for other parts of the country, thus broadening the area of possible spread of the infection. In the country the medical staffs are less prepared to deal with variola cases than those attached to frontier posts, ports or airports.

The disease may also be spread by ignorance, the number of cases multiplying and even occurring in hospitals or among the medical personnel.

As a rule the diagnosis of smallpox is established with a delay of about two weeks but in some cases four weeks had passed before the disease was recognised.

In the 28 foci of imported cases during the 1961-1970 period, 161 of the 391 patients were members of the medical staff or patients admitted to hospital for other diseases during this period.

In the 1967-1970 interval the number of foci caused by imported cases was approximately 400 per year.

A seasonal periodicity has been recorded in the appearance of imported smallpox, two thirds occurring from December to May and coinciding with an increase in the incidence of smallpox in the countries from which it is imported (India, Pakistan). Moreover during this period the spread of the disease appears to be more rapid, for instance in Europe, during the last ten years, the average number of secondary cases was 16 as against 0.4 in the rest of the year.

In Europe most of the imported cases occurred in the northern countries, probably due to a greater exchange with the areas in which smallpox is endemic. Moreover, it was noted that most of the cases were tourists or travellers of the respective countries. Hence, it appeared necessary to survey more closely the immunisation of persons travelling abroad and, in general, to review the vaccination policy (many countries have given up compulsory smallpox vaccination or have used other means: attenuated Rivers vaccine, vaccine inactivated with formaldehyde or ultraviolet light, etc., methods of doubtful efficiency).

During the 1971-1973 interval the onset of smallpox in countries considered as non-endemic, following upon import of the disease from areas with a high endemicity, continued: 4 countries in 1971, twelve in 1972 and four in 1973; in the first half of 1974 this situation affected five countries. In general, the imported cases during this period confirmed previous conclusions; the health authorities were able to restrict the consequences, in keeping with the conditions prevailing in the different countries.

There were two different situations in Europe (Table 1.4):

In England, owing to the measures taken in due time the number of cases was limited to four: one imported from India, (Table 1.4), on the one hand, and on the other, a laboratory technician, and two hospital contacts (drawing attention once again to the possibility of contracting the disease following a minimal contact in hospital).

In Yugoslavia (Table 1.4) the experience was edifying, smallpox being introduced in a country in which the social, economic and religious conditions were added to the surprise factor. It became evident that:

(i) the disease was not recognised and the initial cases and secondary ones spread the disease;

(ii) in-hospital infection which appears to play an important role in the foci of intact countries;

(iii) the deficient immunological background of the population (not vaccinated or vaccinated long ago).

On the other hand, this experience showed:

(i) the importance of an accurate evidence of vaccinations, protection of those leaving for countries with endemic cholera and survey of those coming from endemic regions;

(ii) correct organisation and application of the health regulations in the area will limit the cases in time and space: the epidemics from Yugoslavia started in February and the last of the 167 cases was recorded in April.

Finally, a particular situation is that of India: the recrudescence of smallpox in several states (Bihar, Assam, Uttar Pradesh) constantly supplies Nepal with imported cases (January-June 1974 over 100 cases were recorded).

Apart from the loss of human lives, the import of variola cases into other countries also involves upheaval of social life and serious financial losses; for instance, in the U.S.A. (in 1947) a single imported case generated 12 secondary cases necessitating an intense vaccination campaign in New York that included six million inhabitants. In Yugoslavia the mass vaccination campaign initiated in 1972 included approximately 18 million inhabitants and the survey measures taken affected the life of vast regions.

TABLE 1. 4.

The import of smallpox in Europe 1971–1974

1971	1972	1973	1974 (first semester)
—	Yugoslavia 175 West Germany 1	England 1	—

The import of smallpox still remains a serious problem and demands the taking of severe measures.

1. 3. Smallpox virus and pathogenic mechanism

As vaccinia virus, smallpox virus forms part of a large group of viruses known as poxviruses (in the French literature variola-vaccinia group).

This group includes the largest and more complex viruses of the vertebrates, in whose structure a rectangular mass of deoxyribonucleic acid (DNA) surrounded by several membranes of viral origin can be discerned; the symmetry of the nucleocapside has not yet been fully established.

Some poxviruses are related immunologically by a common internal antigen, but they can be divided into subgroups on the basis of their more specific antigens, their morphology and their natural hosts.

Fenner (1968) and Joklik (1968) recently drew up fairly similar classifications of the viruses belonging to this group (Table 1.5).

TABLE 1. 5.

Classification of poxviruses

Subgroups according to the classification of		Poxviruses included in the subgroups
Joklik	Fenner	
1. Vaccinia	1. Vaccinia	Vaccinia Cowpox Mousepox (ectromelia) Rabbitpox Monkeypox Smallpox Alastrim
2. Myxoma	2. Myxoma	Myxoma Rabbit fibroma Hare fibroma Squirrel fibroma
3. Birdpox	3. Birdpox	Fowlpox Canarypox Turkeypox Pigeonpox
4. Paravaccinia	4. Sheeppox	Sheeppox Goatpox Bovine lumpy skin disease
	5. Orf	Contagious pustular dermatitis (orf) Bovine papular stomatitis Milker's nodule virus (pseudo-cowpox)
5. Viruses not classified	6. Not distributed in the preceding subgroups	Swinepox Fishpox Yaba monkey tumour virus

Some members of the group cause disease in man, animals and birds, usually taking on the form of a generalised infection, with frequent eruptions; some of the viruses are associated with single or multiple tumours of the skin.

The *morphology* of the viruses represents the most important criterion for including a virus in this group, but varies in the subgroups.

In the first three subgroups, considered as true poxviruses, the viral particles measure 200—250 nm/250—320 nm, are rectangular and have an embossed surface.

In the paravaccinia subgroup the viruses are smaller (160—260 nm), more ovoid and their surface appears to be formed of two groups of parallel bands that cross each other at regular intervals.

The *internal structure* presents a central, biconcave mass and lateral bodies that form the common antigen; the whole is surrounded by a coat in which soluble protein antigens are to be found.

The *composition* of the viruses in the vaccinia subgroup includes:

(i) approximately 5% double stranded DNA, with a 80 nm long molecule whose molecular weight is approximately of 200 millions daltons from which GC represents 37%. Having a very great molecular weight, the genome of poxviruses contains sufficient information to codify several hundred proteins;

(ii) lipids represent about 5% of the components (cholesterol, neutral fats and phospholipid whose composition appears to be characteristic of the host cell);

(iii) proteins form 90% and contain up to 18 different components (revealed by immunologic tests). The internal proteins contain specific viral enzymes: DNA-dependant RNA-polymerase; thymidine-kinase, nucleotidphosphohydrolase and several DNA-ases.

In the vaccinia subgroup the viruses could not be differentiated by standard immunologic techniques, but neither could any immunologic resemblance be established between this group and the myxoma subgroup.

Multiplication of the viruses in the variola-vaccinia group takes place only in the cytoplasm; various techniques in cell cultures showed the sequence of viral DNA replication, of viral protein synthesis and maturation of the virion. After adsorption on the cell membrane and penetration into the cell by phagocytosis, decapsidation occurs in two stages.

Cellular infection induces synthesis of the early proteins — among which RNA-polymerases are absolutely necessary for replication—and early and late mRNA. Viral DNA begins to be synthesised 2—3 hours after infection and continues up to the sixth hour.

Assembly of the different parts leads to the formation of new virions that remain within the cell up to the moment of their release by cellular lysis. Maturation and release of the new virus particles begins 7 hours after infection of the cells and continues up to 20 hours.

Of the newly formed genomes only about 30—50% become mature and the released virus particles represent a small part of those produced (48 hours after multiplication ceases only 10% of the vaccine virus formed is released).

During multiplication inclusions characteristic of the virus appear in the cell, type *B* corresponding to the site of viral DNA replication and maturation of the virion, and type *A* inclusions which appear later, do not seem to be at the site of replication and contain only mature particles (they are better considered as a reaction of the cell). Viruses of the vaccinia subgroup do not form type *A* inclusions.

Poxviruses have the capacity to reactivate other inactivated members of the group when present in the same cell; it is a question of metabolic interaction (the release from the denatured coat of the inactivated virus of an intact DNA that can initiate multiplication) and may be followed or not by a genetic recombination (in case of viruses of the same immunologic group — homoreactivation — and respectively, when the viruses belong to immunologically different groups — heteroactivation).

In cell cultures the cytopathic effect (CPE) depends upon the type of cells infected and the nature of the virus. The shape and tinctorial affinity of the cells change and they become detached from the surface of the glass; some rounded infected cells cluster together in foci, for instance, the smallpox lesionin HeLa cells) (Vieuchange *et al.,* 1958).

Joklik (1968) describes two stages in the development of the CPE of pox viruses:

(i) rounding of the cells, which at high multiplicity rates appears within the first hour, representing the response of the cell to ingestion of the particle without any codified information in the viral genome being necessary;

(ii) cellular fusion that begins with DNA replication; fusion appears to depend upon the virus type (occurring more frequently with rabbitpox).

Because poxviruses multiply in the cytoplasm and the viral genome can be isolated from the cell nucleus, it is possible to study them from the genetic point of view. In this way we are able to establish the changes induced by the viral infection and the replication processes of the cell components. Slightly eosinophil intracytoplasmic inclusions appear in the infected cells.

Antigenic structure. In the poxviruses two main antigens have been described: the LS and the NP (nucleoprotein) antigen (Zwartouw *et al.,* 1965).

The corpuscular NP antigen forms an integral part of the viral particle and is responsible for the phenomenon of corpuscular agglutination; it is released by alkaline digestion of the poxviruses and gives a cross reaction with all the sera, therefore representing a basic taxonomic criterion for the classification of any new poxvirus.

The soluble LS antigen has a thermolabile constituent L (destroyed at 60°C) and a thermostable one S (resistant to 100°C). Its molecular weight is about 240 000 daltons; it appears to play a part in flocculation, precipitation and complement fixation (Nicoli and Jolibois, 1954 a).

The LS antigen has several protein components that behave as precipitinogens; hence, in the precipitation reaction, according to the Ouchterlony technique, several bands can be detected.

Complement fixation has been attributed to antigen V, bound to the viral particle. The neutralisation phenomenon seems to depend upon the intact viral particle.

Haemagglutinin appears late in the multiplication cycle of some of the poxviruses and is considered a genetic marker. Separated from the viral particles by centrifugation, it is a pleomorphic particle with a diameter of approximately 50 nm, rich in lipids (74% neutral lipids, the rest, phospholipids) and has a density that varies with the nature of the tissue in which it develops.

Stability of the poxviruses. Despite the presence of a membrane poxviruses are relatively resistant to inactivation by common disinfectants, heat, drying and cold; characteristically inactivated by chloroform, but inactivation by ether is variable.

Smallpox and vaccinia viruses exhibit certain properties that are of particular interest:

(i) Smallpox virus does not multiply at 41°C and vaccinia virus develops just as well at 31.5° C as at 37° C; hence temperature may be used for differentiating the viruses within this group.

(ii) A temperature of 36°C partly attenuates these viruses after 24 hours. Inactivation is produced in 30 min. at 56°C and in 10 minutes at 60°C.

(iii) The viruses are very resistent to dryness, which contributes to the spread of smallpox.

(iv) Both smallpox and vaccinia viruses maintain their virulence at low temperatures (down to −180°C).

(v) Lyophilisation maintains all the properties of the viruses, permitting improvement of the vaccine and its use in warm countries.

(vi) Glycerine is a good preservative of vaccinia virus, especially as it also has an antibacterial action.

(vii) The action of antiseptics is marked but to a lesser extent than in the case of bacteria; formaldehyde (1%) inactivates vaccinia virus in 10 min.

(viii) The pathogenicity of the two viruses for laboratory animals differs and has been taken as a criterion of identification.

The smallpox virus is known to produce two types of disease: variola major or true smallpox, and alastrim (Gordon et *al.*, 1966).

The virulence of the smallpox virus varies considerably. Sarkar *et al.* (1973 a) using several tests (chick embryo mortality, suckling mice mortality, the amount of virus in the chick embryo liver, the proportion of positive serologic reactions) showed that the virulence of the virus isolated from hemorrhagic cases was greater than that of the virus from the current cases. Although not all the Indian authorities agree with Sarkar, some consider that in the same focus and even in the same family strains with a different virulence might be found at a given moment.

The pathogeny of smallpox. We do not yet have sufficient data to establish the evolution of the infectious process up to the outbreak of fever.

By analogy with ectromelia (Fenner's experiments, 1949 clinical and pathological studies in patients and laboratory investigations of vaccinia in rabbits have shown that the virus which has penetrated through the respiratory mucosa rapidly passes into the thoracic lymph nodes and enters the blood circulation for a short interval at the beginning of the incubation period.

It is assumed that the viral particles are swiftly phagocytosed by the cells of the reticuloendothelial system, in which multiplication of the virus takes place during incubation. Finally, viruses massively re-enter the blood flow, together with the onset of fever, initiating the toxaemic phase of the disease characterised by predominant macular rashes, generalised aches and pains, malaise and prostration. Virus spreads to the skin where it multiplies in the epidermal cells and the characteristic eruption follows in 3 to 4 days.

Although these stages have not yet been demonstrated in human smallpox, evidence was found of viraemia in the onset of the disease and its intensity correlated with the gravity of the disease in the patients who do not die; smallpox virus is only isolated from the blood in the first days of the disease but in the severe or lethal forms up to the eighth day.

The morphologic alterations consist in localisation of virus during the onset at the level of the vascular epithelium of the subcutaneous vessels, where perivascular infiltrations occur and haemorrhage is manifested after 3 to 4 days by maculo-papular eruptions. Subsequently, necrosis of the infected epithelial cells takes place and the eruption passes successively through the states of vesiculation, pustulation and later crust formation.

2. The clinical picture

Although smallpox dates back so many centuries its basic clinical features have changed very little. In India the disease has remained the same as was described two thousand years ago or is today in Ricketts' monumental work (1908), nevertheless mass vaccination has produced certain changes; in the epidemics of the last few decades an increasing number of modified (attenuated) forms among successfully vaccinated subjects have been recorded. This has to be taken into account, especially in the countries in which smallpox has been eradicated.

The modified clinical forms differ with regard to the intensity and duration of the symptoms from the known classical forms described in the treatises and which are still encountered today in the course of epidemics (Eremian, 1962).

In the modified forms it is of extreme importance to establish a rapid and correct diagnosis, which is possible today by the rapid laboratory methods. This urgency should be emphasised since many of the recent smallpox epidemics have been caused by the first case that was neither recognised nor diagnosed.

According to Rao (1966), the modified, attenuated and atypical cases, not initially diagnosed, are the cause of the persistence of the endemo-epidemic situation in India.

For a better understanding of the disease, it may be separated, didactically, into the four characteristic stages of eruptive fevers:

(i) incubation;
(ii) initial, invasive, prodromal or pre-eruptive stage;
(iii) eruptive stage;
(iv) convalescence.

Each stage has its epidemiologic, clinical and therapeutical importance and involves the use of different laboratory methods.

2. 1. Incubation period

The time that elapses between penetration of the virus into the respiratory pathways and the first signs of the disease is the incubation period. It is asymptomatic and corresponds to the phase of virus multiplication in the reticulo-endothelial system; it ends at the moment in which the virus passes into the circulation again, marking the clinical beginning of the disease.

The incubation may be shorter (less than 12 days) in intrauterine infection of the foetus and in cases of a portal of entry other than the respiratory one *(variola inoculata,* the portal of entry being the skin where a solution of continuity exists). Longer periods of incubation may be encountered in smallpox contacts treated with chemotherapeutics (Marboran) or specific human immunoglobulin (antivaccinal or antismallpox), immediately after infection.

In general, the incubation is 12 days, but may range from 6 to 21 days. The incubation period is longer or shorter in terms of the infecting dose and degree of resistance of the host at the moment in which he contracted the disease. The higher the infective dose the shorter the incubation; the stronger is the immunity the longer is the incubation period and more attenuated the clinical form.

However, the incubation period is not an absolute criterion of appraisal of the evolution and degree of severity of the disease since many cases have been reported in which, after an incubation of less than 12 days, the evolution was short, the eruptive period curtailed and complete recovery set in.

Resistance to infection may be conferred by contact with the vaccinal virus in vaccinated persons or smallpox virus in case of a previous infection. The loss of resistance is possible due to the intervention of occasional factors, as yet not fully elucidated, or to very high infective doses.

2. 2. Initial prodromal stage

As a rule the onset is sudden, with fever and chills, and in children with convulsions. Fever is never absent in smallpox. According to Indian physicians, the temperature rises to 39 — 40°C and is accompanied by unbearable muscular pains, especially in the lumbar region.

In the prodromal stage fever is probably due to the presence of the virus in the circulation and during this period there are more chances of isolating the virus from the blood.

The second most important symptom after fever are the pains in the lumbar region.

Most patients complain of intense frontal, occipital or generalised headache.

Vomiting may be present and, more rarely, diarrhoea from the eruption on the mucous membrane of the gastrointestinal tract.

The disease frequently brings about hallucinations, delirium or somnolence.

These symptoms, added together, give the impression of an accentuated intoxication, of severe alteration of the patient's general condition. Patients are aware of the gravity of the disease, even in the very severe forms (early haemorrhagic form) and maintain their consciousness up to the very end.

In few cases a generalised rash of short duration (10-12 hours) may appear but it is seldom noted after admission, since most cases reach hospital after the onset of the eruption.

The first stage lasts as a rule 48-72 hours but may sometimes be prolonged up to 90 and even 120 hours, especially in the lethal cases. In general, apart from the cases with a severe, lethal evolution, the symptoms tend to improve and fever falls towards the end of the period of invasion, once the eruption begins to appear.

In the immunised cases, although the symptoms are just as severe at the beginning they gradually lessen at the end of this period.

2. 3. Eruptive stage

With few exceptions this stage is never absent in the evolution of smallpox and draws attention to the diagnosis of the disease. The characteristics of the exanthema make it possible to differentiate smallpox from other eruptive fevers.

The eruption appears three days after the onset of the disease, with a maculous efflorescence first covering the face, then the arms,

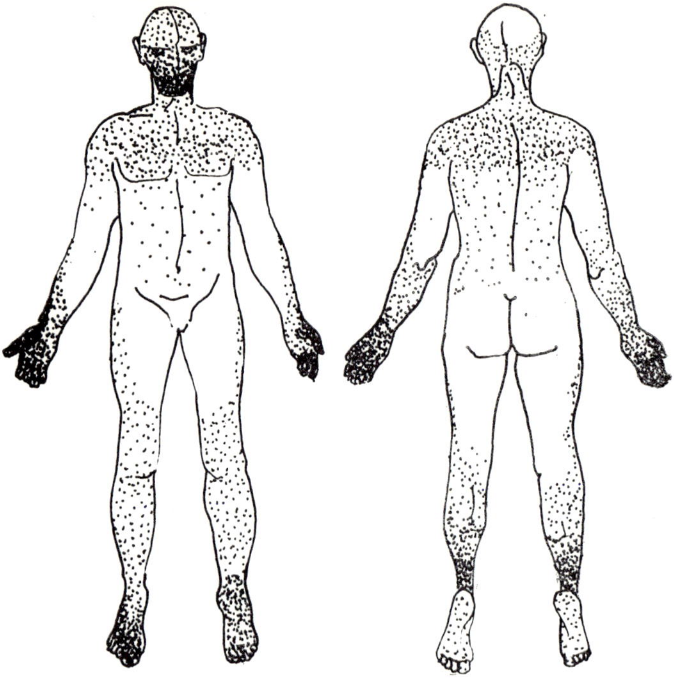

The density and topographic distribution of the eruption in smallpox
FIG. 2.1.

trunk and finally the lower extremities; this is the general order of the exanthema and there are but rare exceptions.

The eruption has a centrifugal distribution (Figs 2.1, 2.2 and 2.3); it is predominant on the face and extremities of the limbs, being more intense on the convex aspects. Similarly, it is denser on the back than on the front of the chest, and less pronounced on the abdomen.

2. 3. 1. Evolution of the exanthema

The eruption passes successively through the states of maculation, papulation, vesiculation, pustulation and later crust formation.

After 36 to 48 hours, the maculae gradually pass into the stage of papules, at first on the face and legs; after two days a papular eruption appears over all parts of the body.

Close examination shows that the papules are centred around a small vesicle which, when pricked with a needle, discharges a slight amount of fluid. According to Rao (1964) this might be used as a simple test for differentiating them from syphilitic papules, should the problem of a secondary syphilis arise.

Gradually the amount of fluid increases, transforming the papules into vesicles.

In contrast to the vesicles of chickenpox, in smallpox the content of the vesicles is turbid and hence the difference between vesicles and pustules is not very great. Their main characteristic is their round, pitted appearance, well delimited by the skin, which differentiates them from the vesicles of chickenpox.

The vesicles may differ in aspect. As a rule they are raised above the skin, are hard, and when touched give the feeling of a glass bead; at times they are much smaller and appear to be sunk into the skin. Finally, they may be very small and few, in number, drying rapidly before becoming pustules.

The monomorphous character of the eruption in smallpox is a decisive element in the differential diagnosis with chickenpox.

After 2 to 3 days' evolution (on the 6th—7th day of the disease) the content of the vesicles becomes turbid (commonly encountered in smallpox) and the vesicles pitted in the centre, i.e. the so called umbilication of the vesicles.

Four to five days later the content is purulent and the vesicles pass into the stage of pustules. This moment is generally marked

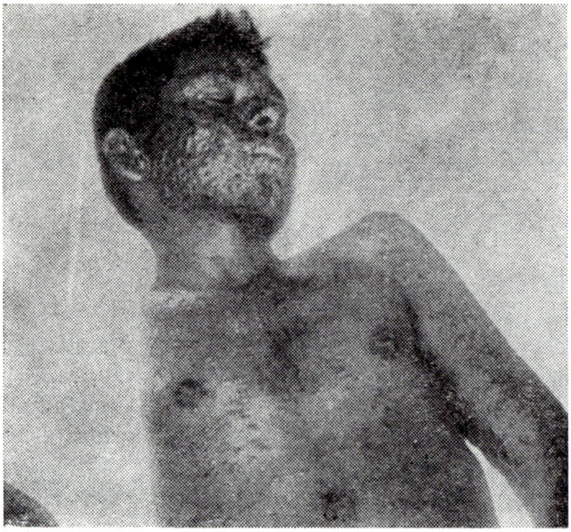

Smallpox, the common clinical form on the 10th day of the disease in a male patient aged 31, vaccinated and revaccinated ten years previously. The eruption has reached the stage of pustulation with a typical distribution: predominant on the face, somewhat less on the upper extremities, rare on the chest and abdomen, absent on the neck (authors' collection).

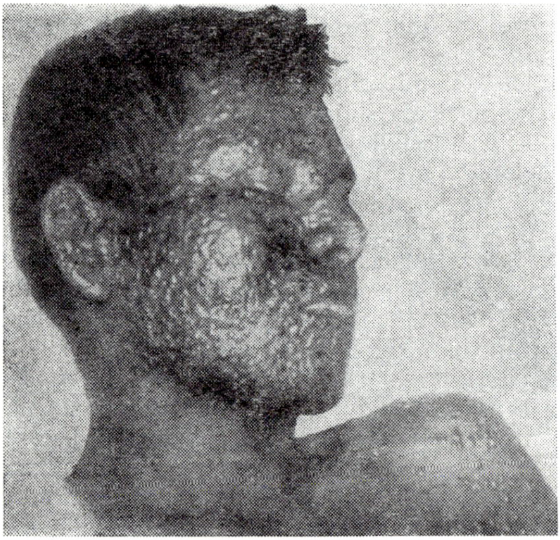

The same case, detailt of the eruption on the face: confluence of the eruptive elements, with haemorrhagic crusts at the level of the commissures, rare pustules on the neck.

FIG. 2.2.

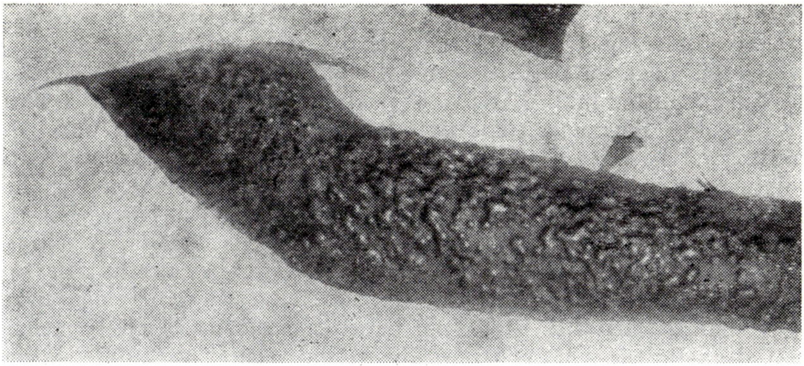

The same case, details of the eruption on the right arm; the eruption is more accentuated towards the extremity of the limb.

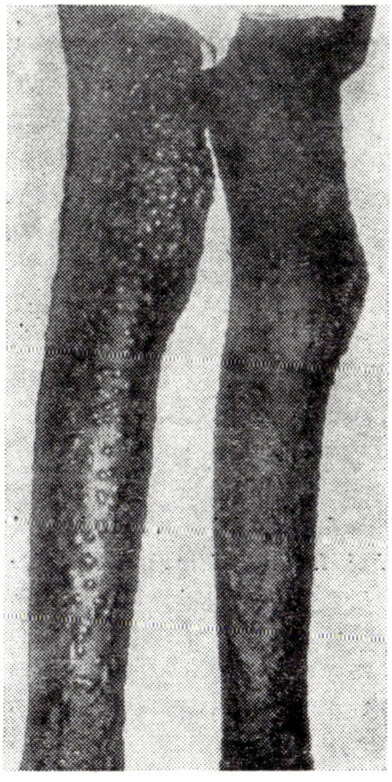

The same case, eruption on the lower extremities: distinct pustules on the anterior aspect of the legs.

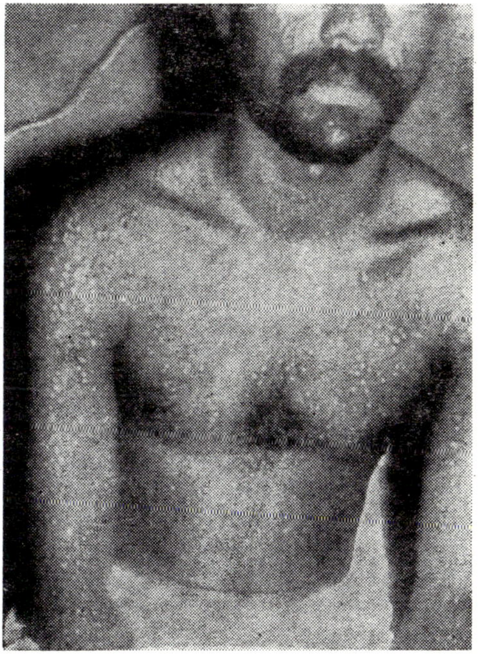

Smallpox in the desquamation stage (authors' collection).

FIG. 2.3.

by a rise in temperature (of shorter duration than in the prodromal stage), producing a state of agitation.

In this stage the patient's general condition is most discouraging; he lies exhausted in bed, refuses to eat and only accepts drink with the greatest reluctance. The entire body is covered with purulent crusts that become confluent in places and below which there is a mixture of pus and blood. At the level of the commissures on the face rhagades appear, suprainfected with bacteria in many cases.

When the patients manage to pass through this stage there follows a period of desquamation (on the 14th—15th day of the disease) which ushers in convalescence.

Exhausted by their suffering, patients gradually begin to manifest an interest in what is happening around them. The pustules begin to dry and fall. Desquamation lasts on the average about 2 to 3 weeks and the patient's face clearly shows the ordeal he has passed through.

After the crusts fall, deep pock marks completely alter the patient's physiognomy so that it is often impossible to recognise him, marking for the rest of their lives those who have managed to survive. When a high degree of immunity exists the exanthema may be more discrete and the pock marks fewer and more superficial.

In the eruptive stage, certain changes may appear that condition a special development, which is sometimes extremely severe and unexpected. In some cases, from the very first day of the exanthema the macules may take on a petechial aspect that develops at the same time as other haemorrhagic phenomena (bleeding from the nose, gums, rectum, etc.); progressive aggravation sets in and the patient dies without the eruption passing through the other stages. Although extremely rare, this early haemorrhagic form is fatal in 100% of the cases.

In other instances, without any apparent cause and irrespective of the moment, the eruption suddenly becomes haemorrhagic, accompanied by the manifestations described. In these cases, too, the course of the disease is extremely severe and the mortality rate almost 100%.

As already shown, the haemorrhagic component of the eruption imprints upon the disease a severe, fatal evolution. However, even in the common forms a congestive halo may be discerned around the eruptive elements, but which does not give to the disease the gravity and fatal course of the haemorrhagic form proper.

By its characteristic aspect the eruption remains a particularly important element both for the diagnosis and for the prognosis of smallpox.

2. 4. Classification

The clinical picture of smallpox (variola major) as encountered in Asia, especially in India and Pakistan, has remained unchanged along general lines down the centuries. However, the initial clinical framework is broader at present and has led several authors to try to systematise the disease according to various criteria.

The first criterion is the eruption.

In terms of the presence or absence of the eruption, smallpox may be separated into two large categories – smallpox with exanthema and smallpox without exanthema. According to the density and topography of the eruption the disease may be separated into confluent, coherent and discrete clinical forms.

Another classification is that based upon the predominance of one of the symptoms. From this point of view, and especially in countries from which the disease has been eradicated, account has to be taken of the forms that differ from the classical description, particularly those modified by vaccination which changes their aspect and usual course, so that it is no longer recognisable even by the best informed specialists.

2. 4. 1. Smallpox without exanthema

In the group of atypical forms a special place is occupied by smallpox without exanthema, *variola sine eruptione*, *variola sine exanthemata*, which may exhibit several clinical forms from the mildest to the most severe.

2.4.1.1 *Pharyngeal form*

The pharyngeal form depends upon the immunity conferred by previous vaccination or revaccination. The local examination reveals moderate hyperaemia of the pharyngeal mucosa, accompanied by a slight difficulty in swallowing.

The eruption itself is limited to a few spots unequally distributed over the soft palate or uvula, or more rarely over the posterior pharyngeal mucosa.

Fever is never absent but is milder in this form, not exceeding 38° to 38.5° C (Fig. 2.4). As the patient's general condition is not much

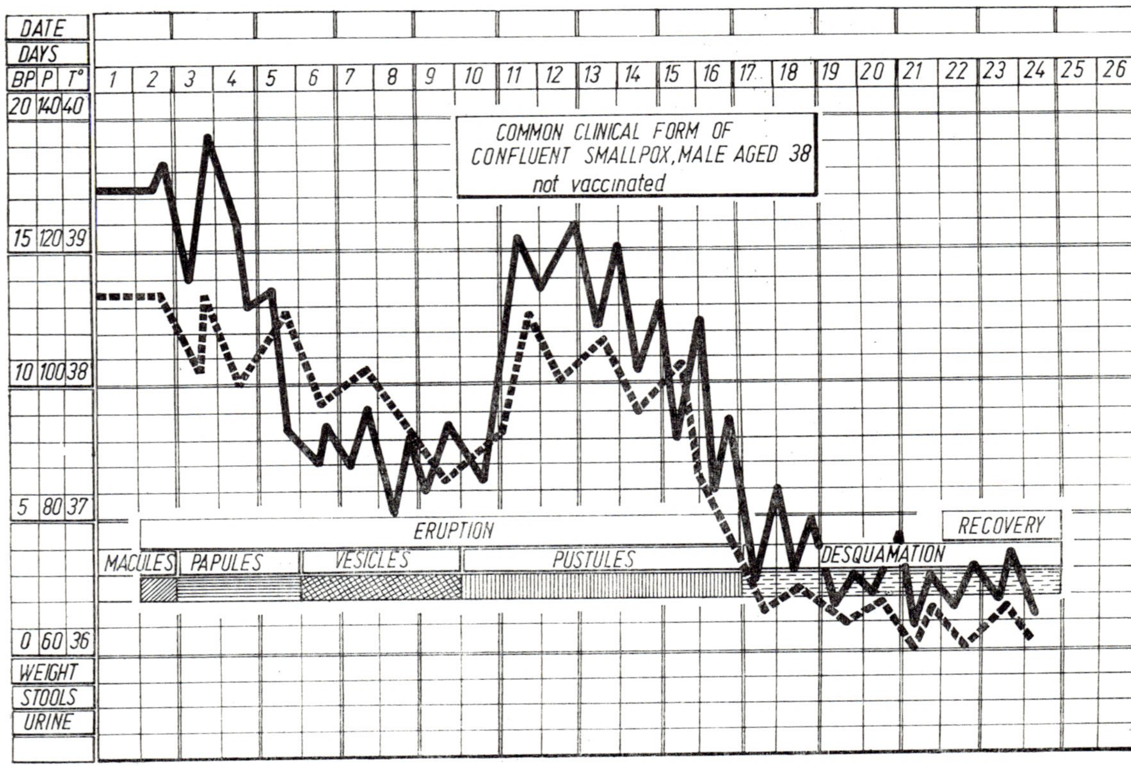
DATE
DAYS
BP P T°
1 2 3 4 5 6 7 8 9 10 11 12 13 14 15 16 17 18 19 20 21 22 23 24 25 26
20 140 40
15 120 39
10 100 38
5 80 37
0 60 36
COMMON CLINICAL FORM OF
CONFLUENT SMALLPOX, MALE AGED 38
not vaccinated
ERUPTION
RECOVERY
MACULES
PAPULES
VESICLES
PUSTULES
DESQUAMATION
WEIGHT
STOOLS
URINE

FIG. 2.4.

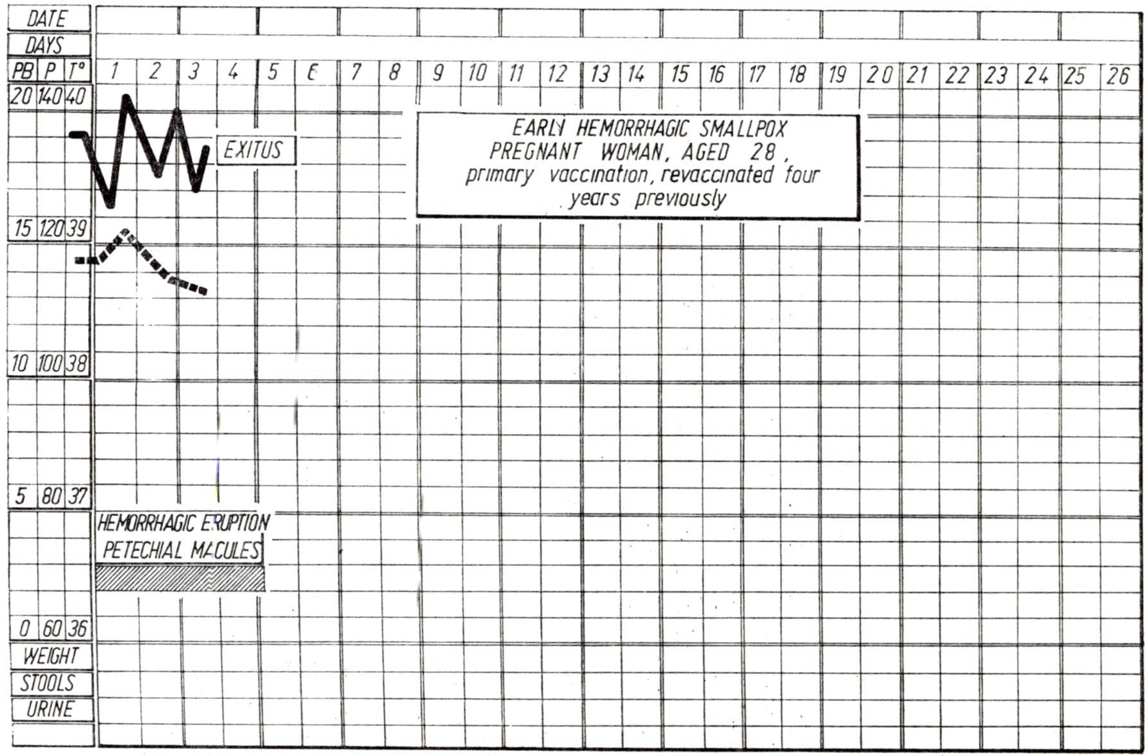
DATE
DAYS
PB P T°
20 140 40
15 120 39
10 100 38
5 80 37
0 60 36
WEIGHT
STOOLS
URINE
1 2 3 4 5 6 7 8 9 10 11 12 13 14 15 16 17 18 19 20 21 22 23 24 25 26
EXITUS
EARLY HEMORRHAGIC SMALLPOX
PREGNANT WOMAN, AGED 28,
primary vaccination, revaccinated four
years previously
HEMORRHAGIC ERUPTION
PETECHIAL MACULES

FIG. 2.5.

altered he continues to go about his work. This is the form contracted by mothers who look after children suffering from smallpox. This form with a mild outcome is frequently encountered in India.

From the epidemiologic point of view the pharyngitis of smallpox represents a great danger because it is not recognised in time, or not at all, and patients come in contact with healthy persons.

In such cases, only a careful epidemiologic survey may lead us to suspect smallpox (the patient may have come from a country in which smallpox is endemic or there is an epidemic, etc.).

2.4.1.2. *The influenza-like form*

The influenza-like form is a rare modality in the evolution of smallpox – as a rule in persons vaccinated and with a low degree of immunity (more than three years after the last vaccination); as in the preceding form it evolves without eruptions.

The onset of the disease starts with the entire suite of characteristic symptoms: fever, pains in the back and lumbar region, headache, nausea, vomiting, etc., but this aspect is limited to the prodromal period, after which the temperature falls and the other symptoms are attenuated. Exceptionally, a transitory rash may appear in the form of macules that may sometimes take on a haemorrhagic aspect.

The symptoms and the absence of an eruption renders this form difficult to distinguish from influenza, especially in an environment of influenza epidemics (Gray, 1949).

2.4.1.3. *The pulmonary form*

Smallpox without eruption may take on a more coherent form with a net pulmonary localisation.

In cases that begin in the same way as the influenza-like forms with characteristic prodromal symptoms, respiratory symptoms develop: coughing, expectoration, dyspnoea, suffocation, cyanosis, fine subcrepitating rales and bronchic rales can be heard. The X-ray shows an uncharacteristic image, with diffuse shadow in both lungs, predominant around the hili and inferior lobes.

The clinical symptoms may be alarming and several authors have described it as the pulmonary form of smallpox (Ratner and Hudeakova, 1961). After a few days the temperature falls and regression of the other symptoms sets in, but the patient is physically exhausted and the convalescence is very long.

2. 4. 2. Smallpox with exanthema

The eruption remains the main clinical criterion of differentiation of the clinical forms, the character of the eruption defining each form in the varied range of the clinical disease.

The aspect of the eruption separates the disease into two distinct clinical groups: *haemorrhagic smallpox* and *non-haemorrhagic smallpox*. As regards their aetiology it is still uncertain whether they are both caused by the same virus or whether the haemorrhagic form is caused by a special type of virus. Actually, the non-haemorrhagic forms have often been found to be the source of infection of haemorrhagic forms and vice versa (Rao, 1966).

This seems to involve a particular immunologic behaviour and also to depend upon the influence of hormonal disturbances, of disorders in the intimate mechanism of formation of the antibodies, the intensity of viraemia, toxaemia, blood disturbances affecting bleeding and clotting, as well as the action of certain unknown factors (Downie *et al.*, 1965; Metha *et al.*, 1967). Recent studies concerning participation of the complementary system and interferon in infections appear to give promising results in this connection.

Many discussions have arisen regarding the delimitation of haemorrhagic from non-haemorrhagic forms. For instance, Rao (1964), who in the course of twenty years treated 25 000 cases of smallpox in the Hospital of Communicable Diseases in Madras, asserts that haemorrhagic smallpox is a definite entity, different from the non-haemorrhagic forms. In opposition, however, Dixon (1962) does not even admit the term of haemorrhagic smallpox since haemorrhages are encountered in very many forms, *in the severe fatal cases but also in very many common cases.*

Even if we do not take into account these considerations and especially the clinical speculations, *the haemorrhagic form remains a distinct and particularly severe form of smallpox*, whose pathogenic mechanism and the factors that determine its unfavourable course, from the beginning or during its evolution, are still unknown.

The haemorrhagic character of the eruption is the decisive element that differentiates the haemorrhagic from the non-haemorrhagic forms.

In the classical treatises the non-haemorrhagic forms of smallpox have been grouped according to the density of the eruption. This criterion cannot be considered as either clinical or of prognostic value. Moreover, this has been recognised by Dixon, although he does not take account of it in his classification. In grouping the non-haemorrhagic forms into categories one should first bear in mind the character of the skin lesions and the density of the eruption.

Rao's classification in which the haemorrhagic form is a distinct entity appears to be more satisfactory for the practical requirements of the physician called upon to care for smallpox patients. According to Rao there are four classical variants: a haemorrhagic variant with two types (early and late) and three non-haemorrhagic variants — flat, common and attenuated. In terms of the density and distribution of the eruption each of these three variants may exhibit three different types — confluent, coherent and discrete. The clinical forms will now be discussed according to Rao's classification.

2.4.2.1. *The haemorrhagic forms of smallpox*

In these clinical forms haemorrhage appears under the skin or mucosa, irrespective of the stage of development of the eruption. This form takes the heaviest toll, the proportion of deaths being higher than in all the other known forms of smallpox. It is more frequent in adults and especially in pregnant women. Smallpox vaccination does not appear to influence the course of this haemorrhagic form, as occurs in the other non-haemorrhagic forms. This also results from the data in Table 2.1.

TABLE 2. 1.

Vaccination status	Morbidity	Mortality
Vaccinated	3.7%	98.3%
Non-vaccinated	2.8%	96.2%

As already mentioned haemorrhagic smallpox presents two distinct types: early and late.

The *early haemorrhagic form* (variola purpura) (Fig. 2.5) presents the highest degree of intoxication. The prodromal symptoms are very severe from the beginning, the period of invasion being as a rule shorter than 3 days.

The patient has a characteristic facial appearance, with bright, burning eyes; he suffers from insomnia and muscular pains, especially along the spine, and a permanent sensation of pressure on the chest.

At the same time numerous haemorrhagic spots appear on the mucosae (especially the conjunctiva).

Death is unexpected, generally on the sixth day when the maculation stage has not yet been reached or even before the appearance of the first eruptive elements. The end is caused by multiple visceral haemorrhage. Specific of this clinical form is its total mortality rate (100%), irrespective of whether the patient was immunised in the past.

TABLE 2. 2.

Vaccination status	Morbidity	Mortality
Vaccinated	1.6%	100%
Non-vaccinated	0.2%	100%

The second, the *late haemorrhagic type* (variola haemorrhagica pustulosa) (Fig. 2.6) is likewise characterised by a high degree of intoxication; the period of invasion does not exceed three days.

Haemorrhage develops after the appearance of the eruptive elements which are as a rule very numerous, flattened or haemorrhagic around their base of implantation. All the symptoms of the disease are very severe, from the onset up to death. The high fever is accompanied by the same lancinating muscular pains. Death occurs after 8 to 12 days.

TABLE 2. 3.

Vaccination status	Morbidity	Mortality
Vaccinated	2.2%	97%
Non-vaccinated	2.6%	96%

2.4.2.2. *The non-haemorrhagic forms of smallpox*

The flat or smooth clinical variant (Fig. 2.7) has an appraisable degree of toxicity. The prodromal stage lasts 3 to 4 days and is characterised by severe symptoms whose gravity is maintained even after the eruption has appeared. The eruptive elements evolve slowly; the flat vesicles give the impression of being buried in the skin, they

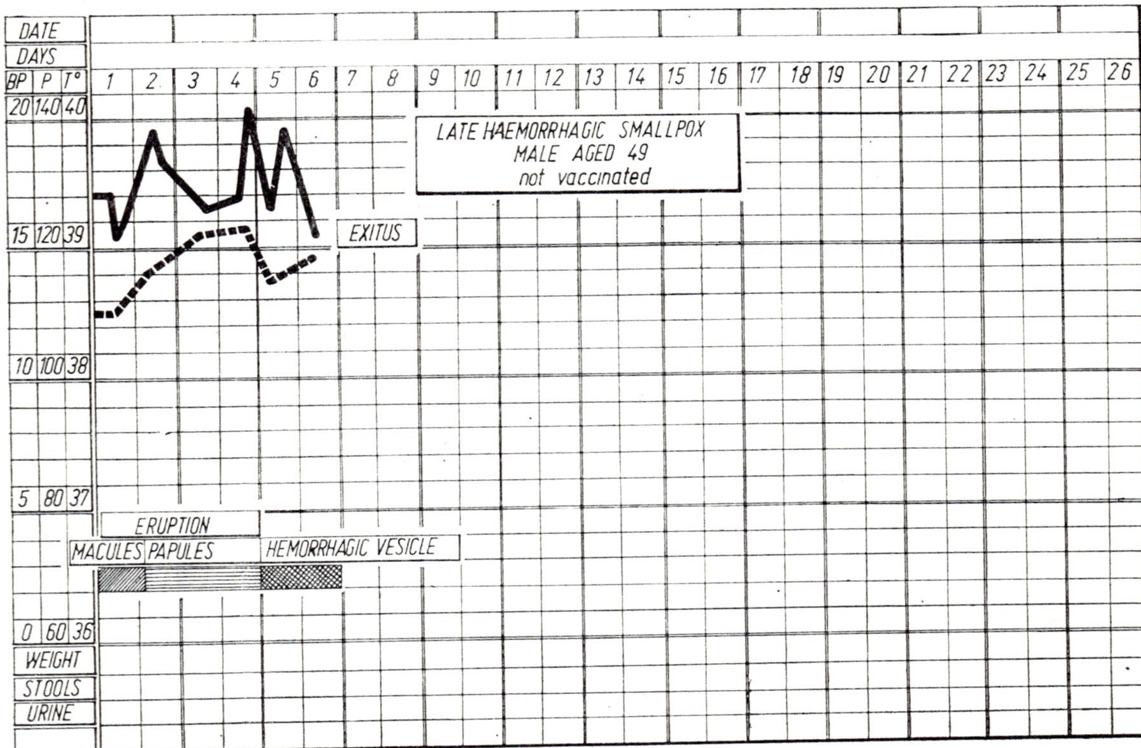
DATE
DAYS
BP P T°
1 2 3 4 5 6 7 8 9 10 11 12 13 14 15 16 17 18 19 20 21 22 23 24 25 26
20 140 40
15 120 39
10 100 38
5 80 37
0 60 36
WEIGHT
STOOLS
URINE
LATE HAEMORRHAGIC SMALLPOX
MALE AGED 49
not vaccinated
EXITUS
ERUPTION
MACULES
PAPULES
HEMORRHAGIC VESICLE

FIG. 2.6.

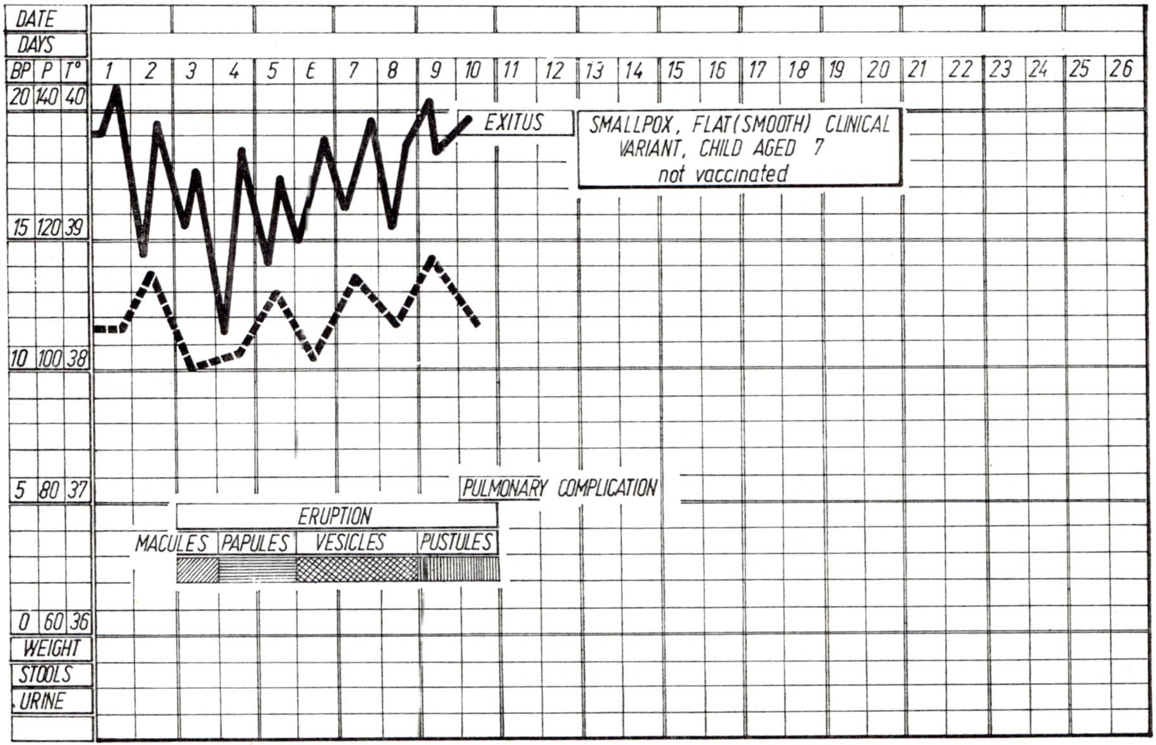
DATE
DAYS
BP P T°
1 2 3 4 5 6 7 8 9 10 11 12 13 14 15 16 17 18 19 20 21 22 23 24 25 26
20 140 40
15 120 39
10 100 38
5 80 37
0 60 36
WEIGHT
STOOLS
URINE
EXITUS
SMALLPOX, FLAT (SMOOTH) CLINICAL VARIANT, CHILD AGED 7 not vaccinated
PULMONARY COMPLICATION
ERUPTION
MACULES
PAPULES
VESICLES
PUSTULES

FIG. 2.7.

contain very little viscous fluid, are not pitted and most of them exhibit haemorrhage at their base and are sometimes surrounded by an erythematous areola.

This clinical variant is more often encountered in children, especially in those that have not been vaccinated, and gives a high mortality rate (see Table 2.4.). In those who survive the pock marks are not very deep.

TABLE 2. 4.

Vaccination status	Morbidity	Mortality
Vaccinated	2.4%	64.5%
Non-vaccinated	10.9%	97%

As regards the density and topographic distribution of the eruption three types may be differentiated:

The first, *confluent type* is characterised by a dense, confluent eruption of the face, extremities and even the back. The symptoms are severe throughout the disease and the temperature ranges between 39° and 40°C. Specific of this form are the pulmonary complications that cannot be influenced by antibiotics. The morbidity and mortality in terms of vaccination are shown in Table 2.5.

TABLE 2. 5.

Vaccination status	Morbidity	Mortality
Vaccinated	1.4%	79.1%
Non-vaccinated	8.2%	99.3%

The second non-haemorrhagic type is the *coherent* one. It is similar to the first but the eruption, confluent on the face is discrete on the limbs. Mortality in terms of vaccination is shown in Table 2.6.

TABLE 2. 6.

Vaccination status	Morbidity	Mortality
Vaccinated	0.3%	—
Non-vaccinated	1.6%	93.5%

Finally, the *discrete, flattened* type is characterised by a discrete eruption over the face and extremities. In such cases the morbidity and mortality rate in terms of vaccination is in Table 2.7.

TABLE 2. 7.

Vaccination status	Morbidity	Mortality
Vaccinated	0.7%	45.4%
Non-vaccinated	1.1%	85%

The *common clinical variant* as its name shows is the most frequently encountered. The prodromal stage is short (2-3 days) and the symptoms of the disease are somewhat milder. After appearance of the eruption the onset symptoms, fever included, abate. The eruptive elements develop and pass through all the stages described. The vesicles, clearly contoured by the skin give, on palpation, the sensation of glass beads. After several days they become pitted, and their content turbid, the vesicles passing into pustules. After a few days the pustules dry and crusts form; when they fall off deep scars remain marking the patient for a lifetime. As shown in Table 2.8. the mortality rate is fairly low among the vaccinated persons and much higher in the non-vaccinated ones.

TABLE 2. 8.

Vaccination status	Morbidity	Mortality
Vaccinated	58.3%	1.4%
Non-vaccinated	82.5%	26.5%

As with the discrete variant, there are types of eruption according to the density and topologic distribution.

The common *confluent type* (Fig. 2.8) has a prodromal stage whose intensity and symptoms recall the classical description of smallpox. Towards the end of the onset period the temperature tends to fall but never below 37°C. The eruption is confluent on the face and extremities. With transformation of the vesicles into pustules the temperature rises and persists up to the end of the desquamation period. Table 2.9 gives the morbidity and mortality rates in terms of vaccination:

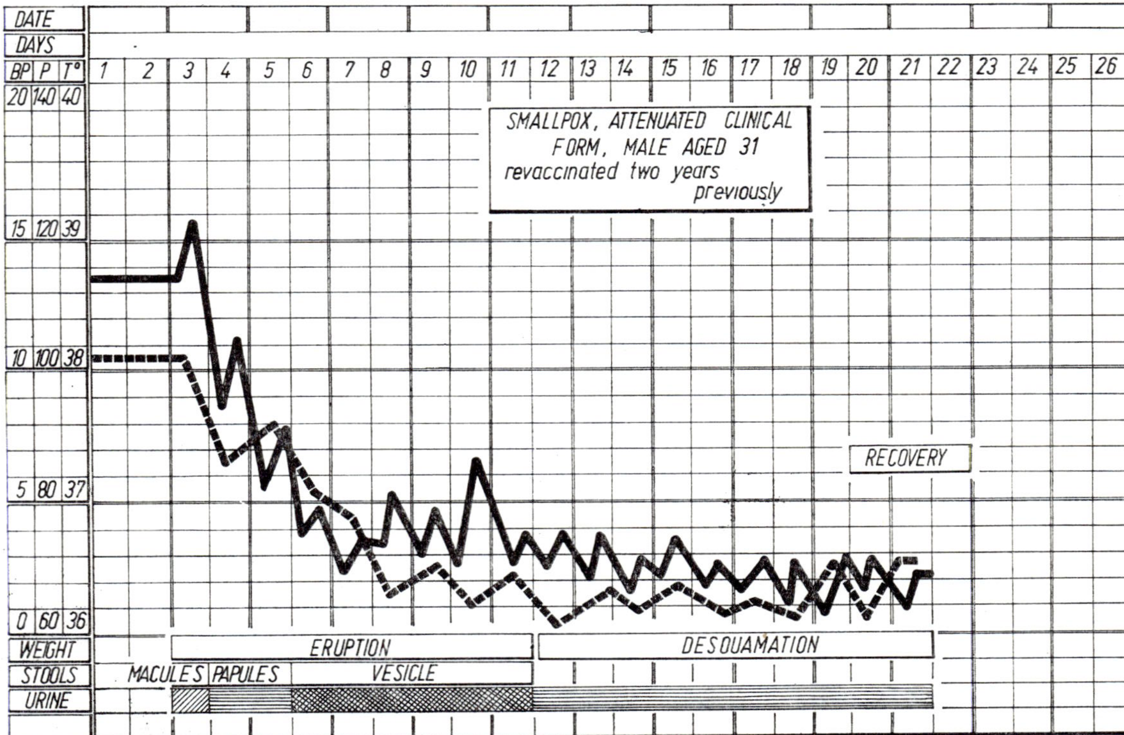

DATE
DAYS
BP P T°
1 2 3 4 5 6 7 8 9 10 11 12 13 14 15 16 17 18 19 20 21 22 23 24 25 26
20 140 40
15 120 39
10 100 38
5 80 37
0 60 36
SMALLPOX, ATTENUATED CLINICAL FORM, MALE AGED 31 revaccinated two years previously
RECOVERY
WEIGHT
STOOLS
URINE
ERUPTION
DESQUAMATION
MACULES
PAPULES
VESICLE

FIG. 2.8.

TABLE 2. 9.

Vaccination status	Morbidity	Mortality
Vaccinated	6.1%	10.2%
Non-vaccinated	28.3%	54.8%

The common *coherent type* exhibits similar symptoms but the rash is confluent on the face and discrete on the extremities. In this form the second bout of fever that marks pustulisation may be absent and the clinical evolution is favourable, especially in the vaccinated subjects, as shown in Table 2.10.

TABLE 2. 10.

Vaccination status	Morbidity	Mortality
Vaccinated	8.6%	1.4%
Non-vaccinated	25.6%	21.1%

In the third common *discrete* type the eruption is mild, the second bout of fever does not appear and the course of the disease is benign, especially in the vaccinated subjects as shown in Table 2.11.

TABLE 2. 11.

Vaccination status	Morbidity	Mortality
Vaccinated	43.5%	0.14%
Non-vaccinated	28.5%	3%

The *attenuated clinical variant* (modified or varioloid) is encountered chiefly in vaccinated subjects. The prodromal stage lasts 2—3 days. The onset symptoms may be attenuated or severe and disappear with the onset of the eruption. The eruptive elements evolve rapidly and desquamation begins early and ends about ten days after the appearance of the exanthema. In most cases there is no second bout of fever and the rash is discrete or even absent.

Here, too, three types may be differentiated according to the density and topographic distribution of the eruption.

Type 1, *attenuated, confluent* is characterised by a confluent eruption on the face and extremities, but does not pass through the stage of pustules. The crusts are thin, desquamation begins soon and the pock marks are superficial. There is no second bout of fever and no lethal cases, as may be seen from Table 2.12.

TABLE 2. 12.

Vaccination status	Morbidity	Mortality
Vaccinated	0.1%	—
Non-vaccinated	—	—

The second *attenuated coherent* type differs from the preceding type by the rash which is confluent on the face but discrete on the extremities. Nor in this type are there any deaths (Table 2.13).

TABLE 2. 13.

Vaccination status	Morbidity	Mortality
Vaccinated	1.9%	—
Non-vaccinated	0.2%	—

Finally in the *attenuated discrete* type the rash is discrete both on the face and the extremities. The morbidity in terms of vaccination is shown in Table 2.14.

TABLE 2. 14.

Vaccination status	Morbidity	Mortality
Vaccinated	32.8%	—
Non-vaccinated	3.7%	—

The persons with a high immunity contract an altogether different clinical form to those described since no eruption appears. The disease begins with the characteristic onset symptoms and is over after 2—3 days. This form is encountered especially in immunised mothers looking after their infants with smallpox.

2. 5. Complications

Smallpox is a very severe disease that kills directly, the contribution of other germs (bacteria, fungi) being in general insignificant. It had been assumed that the second bout of fever in the stage of transition to pustules was caused by suprainfection of the vesicles; this, however, proved to be inaccurate. In the course of the disease accentuated viraemia develops with an extremely severe state of intoxication that exhausts the patient.

This exceptionally severe condition is manifested by a varied range of nervous disturbances, splitting headaches, spinal and lumbar pains, hallucinatory states or delirium, depressive psychosis, melancholia, manic-depressive states, still encountered today in the course of the convalescence of some patients and which may last a fairly long time.

Renal, myocardial, ocular localisations so frequently encountered in the past are more rare today. Although the danger of becoming blind because of the ulcerations on the cornea seldom occurs now it should be taken into account.

Suprainfections with bacteria aggravate the disease which is already so severe. Unfortunately, the complications with the common pyogen germs are not significantly influenced by antibiotic therapy.

The mechanism of the most feared complication, that of haemorrhagic transformation of the eruption, is as yet insufficiently understood.

2. 6. The clinical and epidemiological diagnosis

It should always be borne in mind that the success of the measures taken to arrest an epidemic in a country in which smallpox has been eradicated, depends to a great extent upon the diagnosis of the first case. Therefore everything should be done to establish the diagnosis at an early date — at the level of the first degree contacts when possible. This involves a specialised medical personnel able to corroborate the epidemiologic-anamnestic data rapidly with the clinical findings.

2. 6. 1. The clinical diagnosis

The clinical diagnosis is a valuable element that has often made it possible to detect the first smallpox patient.

On establishing the clinical diagnosis account has to be kept of the inevitable onset with fever, high or low, accompanied by the other symptoms, especially spinal and lumbar pains.

During the onset stage, when the eruption has not yet appeared, the disease may be readily confused with other viral respiratory diseases, especially influenza when there is a concomitant epidemic of the latter. The influenza-like form of smallpox is a clinical reality.

In the prodromal stage, smallpox may also be taken for typhus, states of various aetiology, encephalitis, meningitis, endocarditis septicaemia, acute polyarticular rheumatism, etc.

From the moment in which the eruption appears, the diagnosis can be readily established by its association with the severe, alarming symptoms presented by the patient. The differential diagnosis with infectious diseases in which the eruptive elements reach the stage of vesicles, especially chickenpox and generalis vaccinia, is sometimes fairly difficult. The monomorphous character of the eruption, the circular delimitation of the vesicles and, in general, the entire evolution of the eruption are useful elements for the differential diagnosis.

In the course of pregnancy future mothers may contract the disease even if they have already been vaccinated; in such cases smallpox may be confused with imminence of abortion and toxic pregnancy; several times patients have passed through more than one clinical department (maternity, haematologic or dermatologic services) to be finally referred to the hospital of communicable diseases.

The diagnostic problems raised by haemorrhagic smallpox may begin with the first haemorrhagic manifestations (bleeding from the nose, gums, rectum), with haemorrhagic syndromes accompanied by haematemesis, haemoptysis, petechial eruptions, ecchymosis or various haemorrhagic syndromes followed by perforations, internal haemorrhage or even systemic diseases such as leukaemia, agranulocytosis, etc. In many instances the diagnosis cannot be established during the patient's lifetime but only at necropsy.

As regards the maculous and papulous elements of the eruption, the question may often arise, as observed in India, for instance, of the differential diagnosis with the roseola of secondary syphilis; after a few days the rash in smallpox evolves towards the other stages whereas in syphilis they do not change.

2. 6. 2. The epidemiological diagnosis

In smallpox, *the epidemiologic diagnosis is of capital importance especially in the atypical forms since upon it depends the amplitude and success of the control measures to be taken.*

Of the utmost importance for the diagnosis of the first case is to correlate the clinical symptoms with their onset in a person coming from endemic smallpox regions or a locality with a smallpox epidemic. The disease may be brought into the country by two routes:

(1) By the inhabitants coming from endemic regions or a smallpox area who have obtained vaccination certificates by complaisance. These individuals may be "healthy" on leaving and after one or two days exhibit signs of disease (especially today with the rapid means of air transport). There is no law compelling medical examination of these persons (diplomats, business men, tourists, etc.) and if there were would be needed within the first 12—15 days, which is not practically possible.

(2) Individuals starting from their own country after being vaccinated may return after a few days with fever and with or without an eruption that is not diagnosed (Baroian and Serenko, 1951: the Soviet painter who on returning from Pakistan triggered the smallpox epidemic of winter 1960). As a rule, in vaccinated persons the disease is modified, atypical and attenuated in terms of the degree of immunity conferred by the vaccine. For none of these cases does a law exist providing for quarantine or at least a medical survey within 2—3 weeks after their return.

Therefore, a decisive role is played by the epidemiological survey for a presumptive diagnosis of smallpox.

The survey calls for detailed analysis carried out by the physician, irrespective of his specialty, of all persons with febrile states coming from endemic regions. If the fever is accompanied by eruptive elements, the suspicion becomes an alarm signal and, until the diagnosis is established, all measures of control should be initiated. It should be emphasised that a false alarm is preferable to a smallpox epidemic, especially after the subject has come in contact with a second series of persons, the control measures having to include a far vaster number of contacts extending up to thousands, hundreds of thousands or even millions, and complicating the entire action of prevention.

The epidemiologic state of alarm must be based upon a rapid laboratory diagnosis permitting, together with the clinical and epidemiologic diagnosis, a final one of smallpox.

The laboratory diagnosis is laborious and in terms of the stages of the disease may involve different methods, with varying degrees of rapidity. It may also depend upon the technical facilities of the laboratory specialised in this direction.

In current practice the clinical data have to be corroborated operatively with the epidemiologic findings.

2. 7. Immunity in smallpox

Smallpox induces the appearance of specific antibodies, the neutralising and haemagglutination-inhibition (HAI) antibodies appearing before complement fixation (CF) antibodies (Vichniakov, 1968).

Evidence of neutralising antibodies may be found in vaccinated patients on the third day of the disease and in the non-vaccinated patients on the 5th-6th day. The titres are higher in the former. The appearance of neutralising antibodies coincides with the clinical improvement that usually sets in about the fifth day.

HAI antibodies are detected on the third day almost at the same time as the onset of the eruption, with a rapid increase on the following days. In non-vaccinated patients the appearance of HAI antibodies in a mean titre of 1:160 or more, in the serum sample collected at the beginning of the eruption may be considered as indicative of the diagnosis.

In vaccinated patients CF antibodies can be detected starting from the sixth day and in non-vaccinated patients after the tenth day.

The resistance to a new infection is very strong and a recurrence very seldom occurs (1 in 1000 cases) and only after a long interval (as a rule about 15 years). The disease is generally milder the second time.

However, very severe forms have been noted; for instance, the case of a stretcher bearer in the hospital of Madras, who 15 years after the first time hc fell ill, contracted the haemorrhagic form and died. The disease may likewise recur at short intervals (a case described in India when the patient fell ill a second time only one month after the first).

Those who have suffered from alastrim may contract the major form, but in most cases it is very mild.

2. 8. Treatment

In spite of the progress made during the last few years in the therapy of infectious disease, following the introduction of chemotherapeutical substances and antibiotics, the treatment of smallpox still remains an unsolved problem.

At present we have no efficient chemical drug or antibiotic for treating smallpox once it has become manifest, every attempt having failed.

Great hopes arose when thiosemicarbazone-marboran was found to inhibit growth and multiplication of the vaccinia virus, protecting the mice inoculated by intracerebral route with vaccinia (O'Sullivan *et al.*, 1964). Its therapeutic application did not give, however, the expected results (Bauer and Sadler, 1960; Pickford, 1962; Rao and Mc Fadzean, 1966; Bauer *et al.*, 1969).

Certain results were obtained with specific immunotherapy using smallpox convalescent serum or the serum of subjects recently vaccinated against smallpox, and applying it during the last days of incubation or the first days of the disease. In the experiments carried out in Madras, with specific antivaccinia human immunoglobulins, Kempe (1961) obtained a decrease of the mortality rate of one quarter. These immunoglobulins are, however, very costly and produced in very few countries in the world (among them Romania) so that their use is restricted.

Good results were obtained, according to Soviet authors, with a specific antivaccinia purified animal immunoglobulin (prepared in calves and sheep) used during the epidemic in Moscow, both preventively and therapeutically. Although it has been recommended as being preferable to the specific human immunoglobulin, the serum accidents it causes impairs its advantages (high titre obtained by hyperimmunisation of the animals).

In spite of aetiotropic deficiences, the current symptomatic therapy of smallpox is sufficiently individualised. The treatment applied in the Kasturba Hospital of Infectious Diseases in Bombay (Dr

Desphande) appears to meet satisfactorily with the principal local and general therapeutic desiderata.

Nursing is of particular importance and the greatest attention should be paid to the hygiene of the cavities and mucosa, which are washed with emollient and slightly antiseptic solutions. Of particular importance is the hygiene of the mouth and especially of the eyes to prevent ulceration of the cornea which results in blindness. The skin must likewise be kept clean and washed with weak potassium permanganate solutions.

Locally, depending on the case, the lesions are powdered with sulphonamide talcum powder, or an ointment with antibiotics (streptomycin and penicillin) and hydrocortisone is applied.

It is contra-indicated to administer any medication by the parenteral route. In order to lend support to the defence forces of the organism the patient is given vitamins of the B complex, vitamins C and A, as well as glucose (5%) by intravenous route or in enemas.

During the vesiculation-pustulation period long-acting penicillin is administered (1—2 doses at 7—10 days interval), in order to prevent bacterial suprainfections. In some cases intravenous perfusion with tetramycin associated with corticoids is administered (30—40 mg delta cortisone per day in short 7—12 days series).

During convalescence when desquamation begins, the patients are bathed in a weak potassium permanganate solution and sponged to detach the crusts from the skin.

In the haemorrhagic forms various antihaemorrhagic medications, blood transfusions, specific antismallpox immunoglobulins have been tried without results. Mackenzie (in Madras) recommends the transfusion of frozen or lyophilised fresh plasma or fresh blood, accelerin and vitamin K. The use of fresh plasma appears to be specially indicated in these cases in which thrombocytopenia is constantly present. Heparin is recommended in the cases with intravascular coagulation. Corticotherapy has not proved efficient in the haemorrhagic forms, although it has been recommended. In the 1972 epidemic in Yugoslavia the results obtained were not conclusive. Corticoids must only be administered under the protection of antibiotics. Up to date none of the treatments applied in these exceptionally severe haemorrhagic forms have given any result.

Alimentation of the patients is a delicate problem in which the perseverance and patience of the ancillary staff may work wonders. In the Kasturba Hospital of Bombay and in Madras Hospital the nursing staff competently help the medical staff, proving their devotion and perseverance under conditions in which they risk their lives, working hour by hour at the bedside of smallpox patients. Their humanity is emphasised by the fact that they never give up a case as lost, until all is over.

2. 9. The smallpox hospital

In the countries in which smallpox is endemo-epidemic the patients are admitted to completely isolated premises within the precincts of a hospital for communicable diseases or in special hospitals. In the countries in which smallpox has been eradicated the problem of quarantine, isolation and treatment demands a series of measures of organisation that are not always easily implemented.

At least in the capital city of a country there must be a hospital which at any moment can insure the isolation and treatment of the first smallpox cases. Practically, there should be a special annexe preferably in a hospital for communicable disease, for the quarantine of suspects and isolation and treatment of smallpox cases; when necessary other annexes or even the entire hospital may have to be taken over for smallpox cases, since admission of smallpox suspects to hospital is compulsory. Apart from the annexe for smallpox cases a laboratory (reference laboratory) must be provided, depending upon a competent specialised staff able to establish the laboratory diagnosis as soon as possible. The personnel, of particular importance in the case of an epidemic, must be chosen, instructed and vaccinated in due time so as to be ready at any moment to start work. Confirmation of the aetiologic diagnosis within as short an interval as possible presupposes a well trained organisation in order to be able to meet with the requirements at any stage of the disease (prodromal stage, period of eruption, desquamation phase), the tests being different for each stage. The responsibility for the prevention of spread of smallpox and for coordinating and starting the entire complex activity in case of an epidemic rests with a Public Health Inspector (depending on the Ministry of Health), who, being in a central position, will be able to insure operational efficiency.

Attached to this Inspectorate there should be several consultants who meet periodically to check on the national scheme for the prevention and control of smallpox.

The whole complex action demands a simple, efficient line of communication between inspectorate-hospital-laboratory. It is sufficient to mention that the coordinating body must be able at any moment to communicate with the hospital concerning the place at which the patient is to be found and the means by which he can be transported. These measures cannot be sufficiently emphasised (isolation of the patient, rapid transport in polyethylene bags to the nearest hospital, if possible, and his treatment there up to complete recovery), their application being of paramount importance.

For this purpose:

(i) there must exist a plan of activity for the inspectorate, hospital and specialised institute;

(ii) a test alert should be given that will set in motion the entire organisation, so as to be able to detect defects and correct them. In the course of these test alerts, the persons engaged in this action will acquire experience.

This organisation may be extended and adapted to the various health and administrative units throughout the country. But the essential point must not be overlooked, the existence of a simple rapid communication system between the units and between them and the central control body.

In order to obtain a high efficiency and to instruct the personnel of the various units it will be necessary to:

(i) hold conferences with the medical staff in order to bring them up to date with the clinical aspects and evolution of smallpox, new therapeutical possibilities, and data concerning the management, of epidemics, etc.;

(ii) hold lectures for the population in which competent doctors should describe suggestively and convincingly (without too many theoretical details) the essential elements of prevention and control in smallpox;

(iii) have booklets referring to these aspects;

(iv) have textbooks dealing in detail with the various problems linked to the complex control of smallpox;

(v) have documentary films for the medical and nursing staff;

(vi) have documentary health education films for the masses.

The central authority should draw up a plan providing for:

(i) preventive measures to be applied even when there are no cases of smallpox;

(ii) measures taken when the first case appears;

(iii) measures taken when a case of smallpox appears in the course of natural and social calamities:

(a) earthquakes, floods, drought, etc.;

(b) bacteriologic focuses during campaigns.

Each plan should establish:

(i) the organisations and formations that will take part (hospitals, laboratories, etc.) with all their personnel and equipment;

(ii) the mobile teams of specialists that will have to be sent where necessary;

(iii) the measures to be taken for eventually extending the action by appealing to outsiders (especially women);

(iv) the obligations of the various departments (Ministry of Health, Ministry of Home Affairs, Ministry of the Armed Forces, Ministry of Transport, etc.) and the limits of their authority.

The following are some of the measures taken in a hospital providing for the quarantine of suspects, isolation and treatment of smallpox cases.

First of all the hospital has to be chosen, and it is preferable to have a communicable diseases hospital:

(i) situated at the margin of the town;

(ii) offering facilities for the simultaneous isolation in one or several annexes, which implies a horizontal and not a vertical arrangement.

The precincts of the annexe or hospital should have a corresponding protection perimeter.

The pavilion must provide at least 20 beds and should meet with the following conditions:

(i) have several, and at least two, completely separate entrances;

(ii) have the necessary accommodation for lodging the personnel who can no longer leave once a patient has been admitted;

(iii) be well managed by arranging separate circuits;

(iv) offer sufficient space for storing the linen, utensils and drugs;

(v) have a laundry, autoclave, hot air oven, incinerator and all facilities for preparing and sterilising the instrumentation.

Such an annexe (or hospital) must be able to cope with all the problems raised by confirmed patients, suspects or contacts.

The smallpox patient must be isolated until completely recovered; after the end of desquamation he is no longer contagious.

The suspects are kept in quarantine in a separate annexe in the same hospital for 16 days, after which the confirmed cases are admitted to the smallpox unit and the others discharged. The confirmed patients have to be separated promptly from the suspects since the air-borne contagiousness of smallpox is very great and, practically speaking efficient measures of isolation cannot be taken in the same annexe. The medical and nursing staff working in a quarantine annexe comply with the same rules as those caring for smallpox cases, until the situation becomes clear.

Although, in principle, a hospital for communicable diseases is assigned for smallpox cases, it is advisable to study in time the possibility of using any hospital that meets with the necessary conditions.

Of outstanding importance is the special instruction of the personnel appointed to a smallpox annexe or hospital. This should begin by smallpox vaccination, repeated every two years, as in the third year immunity becomes attenuated and may disappear completely. Vaccination must be done with full understanding of the responsibility involved; there must exist accurate, nominal file cards, from the senior medical officer to the last man on the staff. Those who elude vaccination expose themselves to contagion in an epidemic outbreak (with severe, lethal forms), expose other persons with whom they come in contact and create panic. Those who are not vaccinated or whose situation is not clear must be eliminated from the team.

In smallpox epidemics the doctors and nurses have often paid a heavy tribute readily explained by their direct contact with the patients.

The personnel attached to the smallpox unit or hospital will be instructed in all the details and will have to go over their work demonstratively, under concrete conditions, laying particular emphasis on compliance with the preventive measures, particularly the in-hospital circuits. Each person must know exactly where and how he will work, where the linen is kept, where the utensils, instruments, drugs, etc., are.

They must be acquainted with the entire inventory of the annexe:

(i) where the clean linen is kept, the number and kind of sheets pillow cases, etc., how the linen is brought from the linen room to the patient's bed, when the linen is changed;

(ii) where and how the dirty linen (bedclothes, towels, handkerchiefs, personal linen, etc.) is gathered and treated before being taken to the disinfection point;

(iii) where the crockery and cutlery is kept, who hands it out, how it is washed after the patient has eaten;

(iv) who receives the food, where, and from whom, and how it is transported;

(v) where the cupboards with drugs and instruments are, who takes care of them, what drugs are to be found (current drugs, antibiotics, corticoids, etc.), how the drugs are handed out and how the treatment is applied, where the rest is disposed of (vials, compresses, dressings, etc.).

(vi) where the bedpans, urinals, sputum bowls, wash basins, etc. are kept and how they are disinfected, the place where the protection equipment is kept, etc.

The bath used by the patient, when this is possible, is of special importance since care of the skin given in the bath will limit the spread of virus during the desquamation period: the patient is bathed and sponged daily until all the crusts fall.

The unit should have its own laundry and if this is not possible a mobile laundry should be assigned for this purpose.

The incinerator is indispensable; in it are burnt all the patient's contaminated products that cannot be disinfected by other procedures. In India the patient's clothes are also burnt.

The medical personnel is responsible for collecting the contaminated products in the different stages of the disease (pharyngeal washings, pharyngeal exudate, content of the skin lesions by scraping or puncture, crusts, blood) in view of the laboratory tests. The products collected are packed in special containers, prepared previously, then labelled (name of patient, nature of the product, date, day, hour) and sent to the laboratory by a carefully instructed assistant, and in strict compliance with the rules of prevention.

From the moment in which the patient is admitted to the smallpox annexe the only possibility of communicating with the outside world is the telephone.

2. 10. Measures taken when a patient is notified

The moment a patient is notified, the doctor on duty must open the envelope containing the instructions concerning the first measures to be taken in case of smallpox (these have to be brief and to the point).

He will speak to the director of the hospital on the telephone and the head of the department, who will immediately take measures to evacuate the patients from the annexe assigned to smallpox cases and call up the personnel.

He must then check if they have all come, and, if necessary, he will send the hospital ambulance or car to fetch them.

The doctor will then check the vaccination status of each person, eliminating those whose situation is not clear. If more than a year has passed since the last vaccination, it must be repeated. The members of the staff with pyodermitis, allergic or other diseases in which vaccination is contra-indicated for the moment, will be vaccinated under the protection of specific human antivaccinia or antismallpox immunoglobulin, or when these are not available, under standard immunoglobulin protection. When the risk of vaccination is too great, the person in question will be replaced by another having the same training.

The patients from the annexe assigned to smallpox will be evacuated to other wards, according to their disease (without creating panic).

Entry into the hospital is stopped, and a notice to that effect erected: QUARANTINE—ENTRANCE FORBIDDEN.

The central authority will be notified (for instance the sanitary inspectorate) of the gate through which the ambulance will come in and the direct way to the annexe where the patient will be taken over by the assigned members of the staff (the ambulance must not follow the usual circuit but go directly to the annexe).

At the gate through which the ambulance enters in a road sign will point in the direction of the annexe so that the driver need not ask his way and thus come in contact with others.

The protection perimeter, at least 5 metres from the annexe, is closed and several persons put on duty to prohibit the access of any person without a special permit from the director of the hospital.

On the door of the annexe a notice with the word SMALLPOX will be erected.

In the meantime the personnel mobilised will check the arrangements, accommodation, bed linen, and the inventory in general, the bedpans, urinals, wash basins, dirty linen bags, etc., drugs, instruments (syringes, tongue depressor, clamps and forceps, dressings, drums, receiver for soiled swabs and dressings).

The way the circuits are operated is controlled and all routes of access closed so that each circuit shall be separate.

Conditions vary so much that it is impossible to state with any certainty the precise number and qualifications of staff needed for the smallpox hospital, but the following indicates the bare requirements:

(a) a physician specialising in tropical diseases. If such is not immediately available, a good substitute would be a physician with wide clinical experience who is a good organiser, who has energy and is capable of making decisions, and who commands respect and permits no derogation of rules;

(b) a head nurse with experience, courage and authority;

(c) two or three nurses, devoted, and with good professional qualifications;

(d) one or two assistant nurses;

(e) a dietitian;

(f) a responsible assistant for disinfecting duties;

(g) two ambulancemen;

(h) two watchmen

The patient is recieved the moment the ambulance arrives at the hospital, preferably by the door leading directly to his room, and put to bed. From then on the room is completely isolated until the patient's recovery, communication with the outside world being solely by telephone.

The next step is for the case card to be completed, with the day and hour of his admission and all the epidemiological data. The patient is examined and contaminated swabs, garments, etc. collected, packed and transported to the laboratory for testing and confirming the diagnosis. There are strict rules covering the method of packing and despatching contaminated material, and these must be adhered to. Such measures as are indicated by the patient's condition are taken, and the first drugs administered.

An item of particular importance is the disinfection of the ambulance that brought the patient to the hospital. This must not leave the hospital before it has been thoroughly disinfected.

When there is a single smallpox patient the ambulancemen who take him to hospital become first-degree contacts and are kept in quarantine in a special unit or ward. If, however, they have other patients to convey, they must be revaccinated, as are other personnel, and allowed to continue their activities.

If the number of patients requiring transport is large, two or three teams, or more, will have to work in shifts on the same ambulance. They are all classed as first degree contacts but allowed to carry out their duties for one week after being vaccinated and then revaccinated. During the first seven days of duty they must be closely followed up so that if they become infected they can be admitted to hospital.

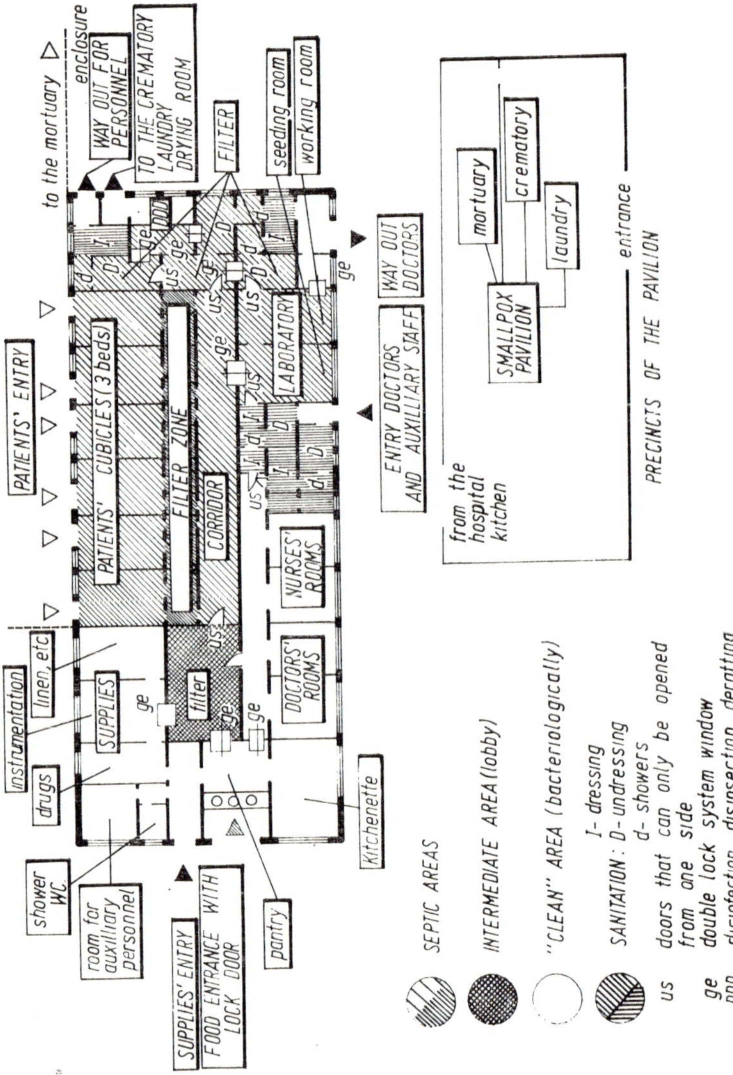

to the mortuary
enclosure
WAY OUT FOR PERSONNEL
TO THE CREMATORY LAUNDRY DRYING ROOM
FILTER
seeding room
working room
PATIENTS' ENTRY
PATIENTS' CUBICLES (3 beds)
FILTER ZONE
CORRIDOR
LABORATORY
NURSES' ROOMS
DOCTORS' ROOMS
WAY OUT DOCTORS
ENTRY DOCTORS AND AUXILLIARY STAFF
instrumentation
linen, etc
drugs
SUPPLIES
filter
shower WC
room for auxiliary personnel
SUPPLIES' ENTRY
FOOD ENTRANCE WITH LOCK DOOR
pantry
Kitchenette
from the hospital kitchen
mortuary
crematory
laundry
SMALLPOX PAVILION
entrance
PRECINCTS OF THE PAVILION
SEPTIC AREAS
INTERMEDIATE AREA (lobby)
"CLEAN" AREA (bacteriologically)
SANITATION: I- dressing
D- undressing
d- showers
us doors that can only be opened from one side
ge double lock system window
DDD disinfection, disinsection, deratting

FIG. 2.9.

The main criterion for the selection of the team working on transport of patients is that they should be vaccinated and fully briefed on the strict routine that must be followed.

A diagram of an ideal pavilion for smallpox patients is shown in Fig. 2.9.

2. 11. Management of the smallpox accommodation

The following data can be considered only as a general outline of management requirements since local conditions will dictate modifications. What needs to be emphasised is that smallpox is an extremely serious airborne disease that is quickly communicated to others through objects that have become contaminated. Every measure must, therefore, be taken to isolate the patient with smallpox so that the premises in which he is housed while being nursed back to health do not become a focus of infection that spreads to other patients and the staff who are caring for them, and to people outside. Even in countries in which smallpox is endemo-epidemic (in India for instance) admission to hospital is compulsory irrespective of the patient's socio-economic or political status, or the clinical form of the infection, which may be so mild as scarcely to require treatment at all.

The ideal is to have a hospital in which all the requirements of quarantine, isolation and treatment of smallpox patients can be carried out. In countries subject to smallpox epidemics such hospitals exist (e.g. the Communicable Diseases Hospital of Kasturba in Bombay) and they have made an outstanding contribution in solving the problem of smallpox. In countries from which the disease has been eradicated the position is altogether different.

For cholera, a hospital is essential, but for smallpox an annexe to a main hospital is sufficient if it is planned to cope with the quarantine aspects and management of such cases. If no annexe or unit is available, any hospital that can be used for the purpose must be utilised provided attention is given to the recommendations laid down.

The requirements are: a ground floor with accommodation for 20 to 25 beds, or a ground floor and first floor with 40 to 50 beds, the latter offering better facilities for quarantine and treatment.

A patient entering the annexe should be admitted by the door nearest to the ward in which he will stay and thereafter that door will remain closed. With small capacity units there is generally a corridor running through the centre of the unit, and the patient's ward should be at one end of it. Next to the door there should be a small lobby or cubicle in which the nursing personnel can hang up their protective apparel, which is worn over their usual hospital clothing, before entering the patient's room.

Protective equipment consists of a gown, tight-fitting at the neck, trousers, headgear (close-fitting bonnet or cap), gauze mask to cover mouth and nose, spectacles, gloves, and foot muffs drawn over the hospital shoes and fastened with laces at the top.

On coming out of the patient's room the protective garments, except for the gauze mask, are removed and placed in a receptacle, which is treated with disinfectant and covered with a sheet of plastic or oil-cloth, and drawn along on a mat soaked with an antiseptic solution to the point (the disinfection point) where the garments are to be disposed of or prepared for further use. Here, the person who has been in the patient's room takes off his mask, washes with soap and water, and takes off his gloves, after having dipped them in the antiseptic solution. He then takes a shower bath, washing his body well with soap and water and giving particular attention to his face and hair.

The lobby or cubicle outside the patient's room is an essential feature, although it may be arranged to serve two rooms.

The disinfection point, also a small lobby or cubicle, must be sited close to the patient's room to avoid the receptacle carrying the discarded garments, etc. having to be moved over long distances.

Food for the patients is prepared in the hospital, not within the annexe, and it is taken into the patient's room by a different route from the one the patient takes when he first enters it. Within this second entry there is a table that has been washed with antiseptic solution, and on this are containers into which the patient's food is placed, care being taken not to touch either the containers or the food. The unit's dietitian carries the food into the pantry, serves the members of the staff, and then the patients. Personnels' dishes and cutlery are disinfected in the lobby and kept in their individual tables. The patient's dishes and cutlery are disinfected in the lobby and kept in his night table, never in the pantry. Nothing the patient uses leaves his room.

It is of the utmost importance that nurses, assistants, dietetian, or whoever is obliged by his duties to enter the patient's room should dress, undress and disinfect their apparel in the same way as the physicians do. Separate routes to his room are arranged for carrying linen, drugs, instruments, etc.

The patient's used bed and body linen are put into receivers with antiseptic solutions and covered with plastic bags and drawn along to the disinfection point on mats.

Attention must be drawn to the role flies play in the transmission of the smallpox virus; especially in summer, all the rooms in the unit must be repeatedly disinfected and wire netting put up at the windows.

Sweeping, and shaking of linen or blankets is strictly forbidden. The floor, walls, beds and night tables must be wiped with a cloth dipped in an antiseptic solution.

Particular attention must be given to pathological specimens (buccopharyngeal and nasal secretions, secretions from the eyes or ears, fluid content of skin lesions, or scabs).

For the staff the emphasis must always be on disinfection and the disposal of the various crusts, secretions and scabs that exude or fall from the patient. In some hospitals in India crusts are gathered up in penicillin vials.

Before discharge, when the convalescent patient is fully recovered and there are no more crusts or purulent secretions, he is given two or three baths with an antiseptic solution, special attention being devoted to hairy parts of the body and particularly the scalp.

Isolation lasts for about 30 to 40 days.

The objects of measures taken within the annexe are twofold:

(1). to isolate and treat the patient;

(2). to prevent members of the public and the staff from contracting the disease and to stop it from spreading to the hospital through the neighbourhood.

A scheme such as that described above is not a fixed one, but should be adapted to suit local conditions.

3. Laboratory diagnosis of smallpox

3. 1. General aspects

The diagnosis of smallpox is based upon corroboration of the clinical, epidemiologic and laboratory data. The importance attributed to this category of data depends strictly upon the epidemiologic conditions in the locality.

In regions where smallpox has an endemic character the identification of new cases does not change the attitude of the specialists in authority; within a characteristic epidemiologic context the clinical manifestations are sufficient to establish the diagnosis. In the countries in which smallpox does not exist or has been eliminated by frontier protection measures and by a high specific resistance of the population consequent on mass smallpox vaccination, the problem takes on an altogether different aspect.

Modern means of travelling, especially by air may help to spread contagion from one area to another by individuals travelling in the incubation period, spreading the disease long before it becomes apparent and has been clinically identified.

Confronted by a typical case, an experienced clinician will be able to recognise the disease. In most cases, however, as demonstrated by the areas recorded during recent years in Europe, smallpox is manifested under different clinical aspects, so that severe forms have been mistaken for purpura, leukaemia, meningitis, pneumonia or even abdominal emergency. The subclinical or abortive forms, especially those with a certain degree of resistance among the populations with a smallpox vaccination programme, represent still a greater danger since they may be confused with other diseases.

A particularly difficult problem is that raised by suspect cases, especially after a recent campaign of smallpox vaccination.

Hence, the situation differs very much from that in countries with endemic smallpox, the laboratory data becoming extremely important for the subsequent measures that have to be taken by the health officials. The results of viral investigations either establish the necessity of taking adequate measures, when a diagnosis of smallpox is confirmed in a patient suspected at first of other eruptive diseases, or when negative avoids the rigorous, difficult, costly epidemiologic measures that upset activity. This implies that laboratory investigations should be done in time for rapid confirmation or rejection of smallpox.

Between these extreme situations are those of the countries with a low smallpox incidence, in which each case is of particular importance and necessitates laboratory confirmation.

Well-equipped laboratories, with good orientation and correspondingly well-instructed personnel can establish a diagnosis of smallpox within a short time. It must not be assumed that all the methods and techniques used in virology for the study of smallpox virus have to be applied, but only the most sensitive and rapid.

At present the WHO recommends the use of a network of laboratories distributed so as to serve several countries; with a minimal number of tests they will give the necessary information for diagnosing smallpox.

Within 24 hours after the samples arrive at the laboratory, it is possible to establish whether it is a poxvirus or another virus that causes eruptives diseases of lesser epidemiologic importance (herpes virus, varicella virus). The products will be examined by the microscope (light microscopy, immuflnouorescence microscopy and electron microscopy) and tested by precipitation in gel and other serologic reactions (flocculation complement fixation, etc.). These procedures are quick and furnish an *orientative* diagnosis.

The latter is completed by additional methods, that *confirm* the diagnosis when smallpox virus is isolated; of the many known procedures, inoculation of embryonate eggs continues to be very useful.

The diagnosis may be confirmed in 2-3 days, the test not only differentiating the smallpox virus from the herpes and varicella viruses but also from other viruses of the pox group (vaccinia or cowpox virus).

The results obtained are sufficient for instituting the epidemiologic measures. It is often necessary to *determine* the type of virus present, i.e. if it is the virus of true smallpox or of alastrim (variola minor).

This diagnosis demands a longer interval of 5-10 days, when the so called *slow* procedures must be used. This may also be obtained at the same time as confirmation of the smallpox infection when the respective techniques are combined. Identification of the two types of smallpox virus is based upon the difference in the maximum temperatures at which they develop, the different mortality rate of the chick embryos after inoculation with various virus concentrations and the results of scarification on the rabbit skin.

The diagnosis of smallpox is based upon identification of the virus or its effects; in view of the emergency of a diagnosis in this disease the serologic tests are useless, since the antibodies appear much later. Determination of the smallpox antibody titres is necessary only for assessing the degree of immunity within a vaccinated population.

In principle, the objects of laboratory tests are to determine the presence of viral particles (microscopic examination) or of the smallpox virus antigens at the level of the skin lesions and in the blood, and isolation and identification of the virus.

What are the methods used for these purposes?

1. *Evidence of viral particles in the smears prepared from the material collected from the skin lesions or mucosa and the blood.* The virus may be observed and identified by several methods (Table 3.1).

(a) *Simple staining* of the smears by the modified Morozov technique; the most recommended is that of Gispen (Lépine preferred the method of Nicolau with phenic isamine).

Simple staining is indicated when the presence of smallpox in the respective area is known and the laboratory test is demanded for confirming each new case; performed with skill it permits a correct diagnosis in almost all the cases examined in time and even differentiation of the vaccinia subgroup from those of varicella or herpes.

This method is sufficient in the early stages of the disease up to the sixth day when the eruption passes into the papulo-vesicular stage. The viral particles that appear in the form of small stained spots, clearly discerned above the limit of visibility, with a characteristic and uniform morphology, represent the diagnostic elements. In the smears from the chickenpox cases the particles are finer, more poorly stained and hardly discernable.

These findings permit a positive diagnosis. A negative result must not however be accepted as such, it must be decided on the clinical and epidemiologic information available.

Table 3.1.

Laboratory methods for the diagnosis of smallpox and differentiation with other viruses

Procedure	Interval	Methods	Differentiation between
Rapid procedures — orientative	24 hours	Light microscope Immunofluorescence Electron microscopy Precipitation in gel Counterimmunoelectrophoresis CF	variola-vaccinia group herpes virus varicella virus
Semirapid procedures — confirmation	2—3 days	Inoculation on CAM Inoculation on cell cultures Inoculation to animals	variola virus vaccinia v. cowpox virus herpes virus varicella v.
Slow procedures — typing of virus	5—10 days	Helbert test Dumbell test	variola major virus variola minor virus

In the subsequent stages of the eruption, this method can no longer be used since on the smears prepared with the material collected from the pustules or crusts, the stained elements that can be seen represent either contaminated microbial flora (especially the cocci produce confusion) or the fragments of destroyed cells.

The presence of these additional elements may result either in a false positive or false negative diagnosis because, although the viral particles are present, they are masked by uncharacteristic particles or contaminating germs.

(b) Staining by *immunofluorescence* has been frequently studied during the last few years in order to establish its advantages and limitations. A response may be obtained within a few hours with immunofluorescence, which has the advantage that the virus can be seen and identified.

The immunofluorescence method demands skill and experience and modern equipment. It is just as specific as the other serologic methods applied to the poxviruses. It is simple, rapid and may be used even when there is insufficient pathologic material for the other

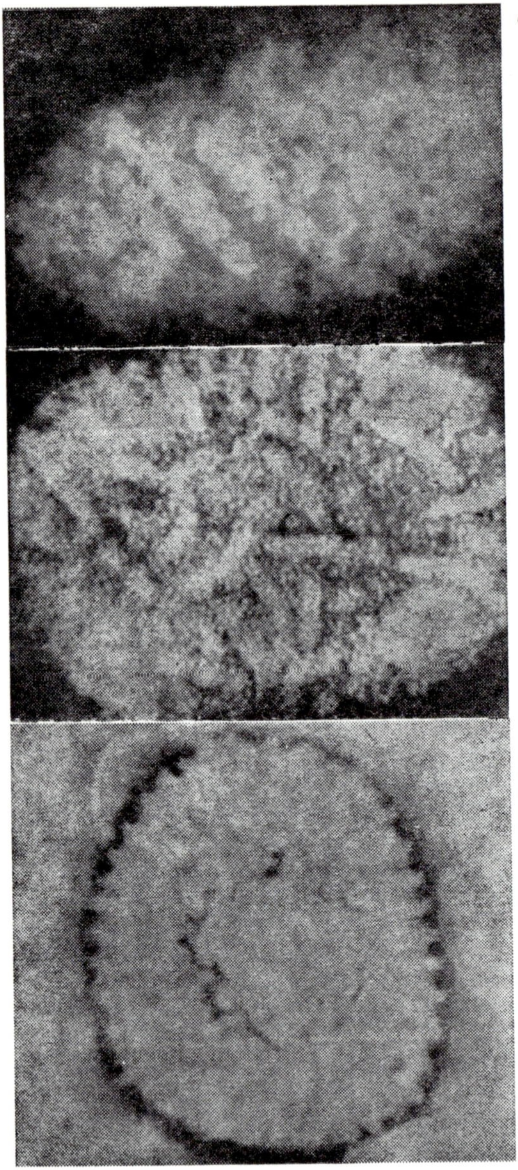

a *b* *c*

Viral particles examined in the electron microscope after impregnation with phototungstic acid: *a* — vaccinia virus; *b* — paravaccinia virus. *Brit. Med. Bull., 1962, 18, p. 193; c*—molluscum contagiosum. *J. Cell. Biol., 1963, 18, p. 51.*

FIG. 3.1.

tests. The products may be fixed when collected and then sent to the laboratory.

However, the method has certain drawbacks, because the preparations may have unspecific fluorescence especially with older products; similarly, cellular retraction sometimes prevents location of virus. Moreover, differentiation between the smallpox virus and vaccinia virus is almost impossible.

As underlined by Grist (1968), because of these disadvantages immunofluorescence cannot be considered as a basic method for the diagnosis of sporadic cases. It is, however, very useful for the rapid screening of numerous samples in a community in which smallpox is endemic or has caused a known epidemic.

Immunofluorescence, direct staining, and complement staining have been recommended, among others, by Avakian *et al.* (1961), Murray (1963), El-Ganzoury (1967), Benda *et al.* (1968) (1968) who combines immunofluorescence with concentration of virus on immunosorbents.

(c) *Electron microscopy* is now considered one of the most rapid and certain methods for the diagnosis of smallpox (Nagler and Rake, 1948; Rake *et al.*, 1948) (Fig. 3.1).

The introduction of electron microscopy in the routine examination for smallpox is based on the sharply different size and aspect of poxvirus elementary particles compared with those of herpes, zona zoster and varicella viruses; the former have a rectangular, bricklike aspect and are much larger than the latter, which are round.

During most of the stages of eruption in smallpox, and, certainly in the early stage in chickenpox, one can usually, but not always, find a sufficient number of viral particles that can be seen by negative staining with phototungsten acid.

Negative staining reveals not only the shape and size of the particles but also the structure of the viral capside.

Nevertheless, especially in the early stage of the eruption, examination by electron microscope may give false negative results probably because of the small number of viral elements, although in these cases viruses may be isolated on the chorioallantoic membrane of the embryonate egg.

Even if the first examination is negative a second examination next day makes electron microscopy one of the most rapid procedures.

The examination proper takes a few minutes but can only be done in appropriate centres; however, a good organisation will guarantee the collection and rapid transport of the samples.

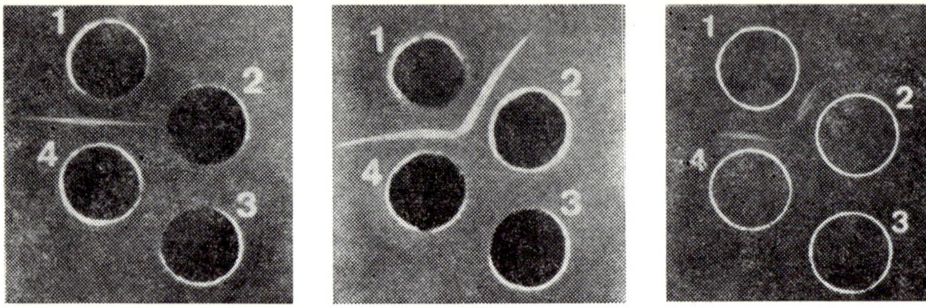

a *b* *c*

Gel precipitation: (a) negative reaction, product from a varicella patient; (b) positive reaction, the united precipitation lines indicate the identity of the pathologic and control sample antigens; (c) presumptively positive reaction, the precipitation lines have not yet joined together.
Distribution of the reagents: 1 — immune variola or vaccinia antiserum; 2 — specimen from lesion of suspect patient; 3 — normal serum; 4 — vaccinia antigen or positive pathologic sample used as control.
(WHO "The Laboratory Diagnosis of Smallpox", 1969).

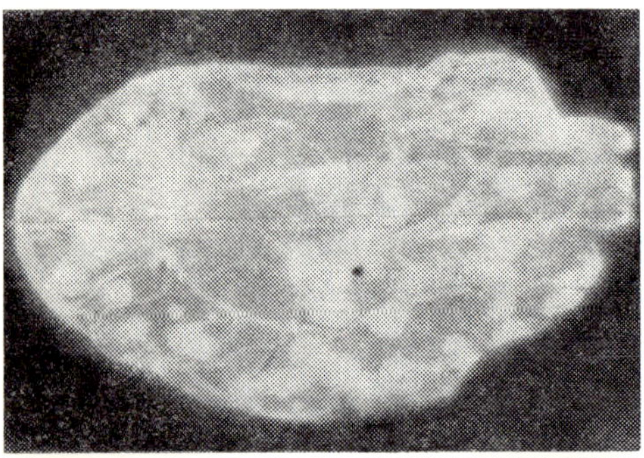

Lesion on the chorioallantoic membrane three days after inoculation of a vaccinia virus suspension (authors' collection).

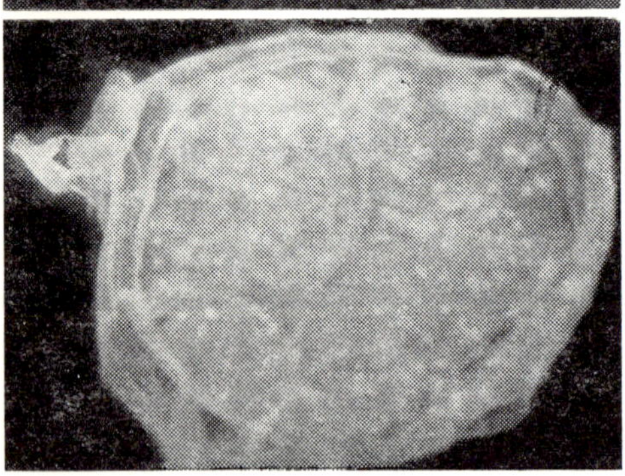

Aspect of lesions on the chorioallantoic membrane three days after inoculation of the material collected from the cutaneous lesions of a smallpox case (authors' collection).

FIG. 3.2.

In the imported smallpox cases in Europe in 1970 in West Germany and Denmark, the laboratory diagnosis was first established by electron microscopy.

2. *Determination of the viral antigens.* The possibility of revealing the viral antigens in pathologic products has led to the development of several adequate methods for the diagnosis of smallpox.

Viral antigens may be detected at any stage of the disease, the method is of particular value in the late stages when visualisation of the virus particles in the smears becomes very difficult.

The smallpox virus may be recognised in the blood towards the end of the incubation period, in the course of the intense viremia that occurs in the severe forms of variola major ; detection of the antigen in the blood is of particular prognostic value, most cases in this category being fatal.

The most adequate techniques for this purpose are *flocculation* or *complement fixation*; the final result is obtained in 18 to 24 hours.

At present the use of *precipitation in gel* is recommended, modified according to Ouchterlony, who uses the slide technique (Marennikova and Malteva, 1961 ; Nicoli and Jolibois, 1964 b) (Fig. 3.2).

Precipitation in gel is a simple method that does not necessitate any particular materials and offers a correct result within a few hours. It has been consistently recommended by the WHO experts.

Precipitation in gel has an advantage over tube flocculation — which involves prior clarification of the product used as antigen — and over the Gispen precipitation method (in a test tube, double unidimensional diffusion) since it is performed with concomitant positive and negative control samples.

By *counter-immunoelectrophoresis* (immunoelectrophoresis, immunc electrodiffusion, electrosyneresis), a technique that combines simultaneous migration in an electric field of the antigen and antibodies with precipitation in gel, recommended by Netter and Sibilat (1968) and Babeş and Kimishkian (1972), a positive result may be obtained within 30 minutes to 2 hours, with a sensitivity four to ten times greater than that of precipitation in gel.

Among the tests used for revealing the viral antigen some researchers include microscope examination by immunofluorescence, which we have presented as an examination for the visualisation and identification of the virus itself.

3. *Isolation and identification of the smallpox virus.* Isolation of the virus remains the most important diagnostic method in smallpox, even when the presence of the virus or its antigens has been made evident by the techniques described.

Isolation of the virus — confirming and determining the virus type — is done by inoculation of the products collected in the early stages of the disease (even in the course of incubation) until the last crust has fallen.

Various procedures have been recommended for isolation: inoculation of the products collected from the patient to *sensitive animals*, inoculation in *cell cultures* and — that which still gives the most satisfactory results — inoculation on the *choriollantoic membrane* (CAM) of the embryonate egg.

On the third day after inoculation on the CAM, the lesions (colonies, pocks) produced by pox viruses (smallpox, vaccinia, cowpox) may be differentiated from those produced by herpes virus; the viruses of chickenpox, zona zoster and measles do not produce such lesions on the chorioallantoic membrane.

Smallpox virus can be isolated under good conditions in *cell cultures*, the cytopathic effect appearing after 48 hours. However, the results differ with the type of cells used and differentiation between the members of the vaccinia subgroup is not so rapid as following inoculation on the CAM.

The virus isolated in cell cultures or on the CAM is identified by serologic tests (hemagglutination-inhibition, complement fixation or precipitation) *in vitro* or by neutralisation tests (in cell cultures or on the CAM), using a smallpox or vaccinia antiserum.

In the vaccinia subgroup the isolated virus can also be identified by determining the maximum temperature at which the virus multiplies, its pathogenicity for the chick embryo and the lesions produced by inoculation to animals (especially on the rabbit skin).

Giant multinucleate cells are usually present in smears prepared from skin lesions in herpes, chicken pox or zona zoster; they do not appear in any other vesicular eruption, therefore nor in smallpox.

This additional test is only orientative and of greater value in the presence of multinucleate cells than in their absence.

3. 2. Collection of specimens for laboratory tests in smallpox

Laboratory examination has as its object to provide evidence of the virus in the samples examined in the microscope, or evidence of the viral antigens by immunologic techniques, or finally, isolation and identification of the virus (see Table 3.2).

The first to be examined are samples from the skin lesions and mucosa; the blood samples are used for confirming the presence of the virus and its antigens in the first days of the disease and for following up the immunologic changes in subsequent periods.

The quality of the material used influences the value of the laboratory diagnosis; the best results are obtained when the samples are fresh and collected by the laboratory personnel instructed in this respect.

Spread of the disease must be strictly avoided during collection, transport and processing of the material.

1. *Collection of the samples from skin lesions in the prevesicular stage.* Samples from the initial lesions of smallpox – macules and papules – are obtained by scraping with the point of a bistoury, or the tip of a Hagedorn needle (generally recommended) or even the bevelled tip of ordinary syringe needles.

As a rule the lesion is not cleaned but some, for instance Kempe and St. Vincent, 1964, recommend wiping it with a cotton swab dipped in ether or alcohol.

Scraping is done so as to obtain as ufficient amount of material from the basement layer of the epidermis, without however drawing blood.

The material is then smeared on a glass slide over a surface of about 1 cm in diameter. Slides should be smeared only once since rubbing can damage the cells and render intepretation in the microscope very difficult.

The samples should be collected from at least six lesions. Several smears from different sites may be prepared on the same slide. In general, 4 to 6 slides are prepared; two of them are stained and the others are used for preparing the suspensions necessary for the other investigations and tests (electron microscopy precipitation in gel, inoculation of embryonate eggs, etc.).

The smears are left to dry in the air without using any physical or chemical means of fixation since they might inactivate or prevent isolation of the virus. Thus prepared the smears may be used for all the tests mentioned.

Once prepared the slides must not be stacked one on top of another but separated by pieces of string, rubber, cardboard, etc.

For examination in the microscope the smears are prepared according to the Van Rooyen technique: the slide is applied on the scraped lesion at several points.

The scraped product may also be introduced into sterile test tubes marking on the outside the place where it was put, so as to facilitate laboratory processing.

2. *Collection of the vesicular specimens.* Both the vesicular fluid and the material obtained by scraping the base of the vesicles are collected.

TABLE 3.2.

Pathologic products used for the laboratory diagnosis of smallpox per stages of the disease

Stage of the disease	Pathologic products	Viral particles	Viral antigen	Virus isolation	Antibodies
End of incubation (in full epidemic)	Blood		±		
	Saliva, nasopharyngeal exudate			±	
	Conjunctival secretion		±	±	
Febrile, prodromal stage (in full epidemic, contacts, suspects from endemic smallpox areas)	Blood		±		
	Saliva, nasopharyngeal exudate			+	
	Conjunctival secretion			+	
Maculo-papular eruption stage	Blood		±	+	±
	Saliva, nasopharyngeal exudate, conjunctival secretion		±	+	
	Maculo-papular fluid	+	+	+	
Vesicular or pustular eruption	Blood			+	+
	Vesicular or pustular fluid	+	+	+	
	Scrapings from the lesions	+	+	+	
Falling of scabs	Blood				+
	Crusts		+	+	
Convalescence	Blood				+

The vesicular fluid is rich in viral particles and antigen and is, therefore, particularly useful for isolation of virus and precipitation in gel.

The lesions are then wiped with a cotton swab moistened in ether or alcohol.

Collection of the vesicular fluid is done with:

(i) small capillary tubes;

(ii) a finely tapered Pasteur pipette;

(iii) needles adapted to a tuberculin syringe.

In the case of capillary tubes and pipettes the fluid is aspirated leaving a small air bubble at the extremity which is then sealed over a flame; the tubes can be stoppered with plasticine. The fluids collected can also be introduced into a single sterile test tube that may eventually contain a small amount of broth or buffer solution.

With the tuberculin syringe, the fluid from several vesicles can be aspirated and the syringe sent with its contents to the laboratory.

Another means of transporting vesicular fluid consists of spreading and drying it on sterile glass slides.

Before scraping, the top of the vesicle is cut off with scissors and introduced into test tubes or vials. Scrapings are then taken from the base of the lesion or smears prepared as already shown.

3. *Collection of specimens from pustules.* After wiping the lesion the tip of the pustule is detached with scissors and introduced into adequate sterile containers; the purulent content is collected with a Pasteur pipette, the tip of a needle or a bistoury. The material collected is introduced into sterile test tubes, where it dries on the walls, or together with the scrapings from the bottom of the pustules, or separately, it is smeared on slides.

Because of the artefacts produced by the degenerated cells the material collected from the pustules is no longer suitable for examination in the light microscope or by immunofluorescence and can only be studied in the electron microscope. This material may be used especially for isolation of virus and detection of the viral antigen.

4. *Collection of the crusts.* In an advanced stage the crusts that form after the pustules dry are detached with fine forceps, the bistoury or the tip of a Hagedorn needle and introduced into sterile tubes or vials, at least 12 crusts in the same container.

During the last stages of the disease the patient may have very few crusts and they should be looked for in his bedlinen.

The crusts cannot be used for morphologic investigations and all attempts at isolating the virus from them have failed. They are used especially for preparing the antigens necessary for precipitation in gel.

5. *Collection of the blood* can be done at any stage of the disease; in the early stage and sometimes even up to the 8th day virus can be isolated from the blood, and in the end stages blood may be used for the serologic tests.

The blood must be withdrawn under sterile conditions so that the person collecting it shall not be contaminated. Moreover, as far as possible, contact of the blood should not come in contact with the content of the skin lesions, the skin being punctured in a place free of eruption.

After withdrawal, the blood is introduced into a test tube with care so as to avoid spreading infective particles; the tubes are sealed with cork stoppers covered in tin foil or with rubber stoppers.

It is preferable to wait for separation of the serum from the clot after collecting and to transport them separately.

6. *Saliva* is collected in 1 ml amounts using sterile flasks.

7. *The pharyngeal or conjunctival exudate* is collected with cotton swabs fixed to a short stiff wire or stick and placed inside the test tubes containing a small amount of sterile broth. This product serves to isolate the virus in embryonate eggs or cell cultures.

The use of these products is recommended particularly when inoculation can be performed swiftly in a nearby laboratory.

8. *Pharyngeal washings* may be collected for the same purpose, taking all precautions to see that the person performing the washing should not be contaminated.

Saliva, the pharyngeal or conjunctival exudate and pharyngeal washing serve to isolate the virus, especially when the eruption is in the stage of maculation or papulation, as well as when marked affection of the upper respiratory tract is observed.

●

Because of the exceptional conditions under which collection of the products is made an adequate laboratory kit has been devised.

At present a kit recommended by Grist (1968) and furnished by the WHO contains:

(i) a Hagedorn needle or vaccination lancet in a stoppered test tube. It is used for scraping the macules and papules, opening them and scraping the bottom or detaching the crusts;

(ii) six clean glass slides each covered in tin foil;

(iii) a container with six capillary tubes (WHO specialists recommend the use of sterile Pasteur pipettes fitted with a rubber dropper) for collecting the fluid from the vesicles and pustules;

(iv) sealed sterile flasks (with a rubber stopper or screw on lid) for collecting the crusts or fluid from the vesicles and pustules.

The container of wood or metal has double walls or two boxes are introduced one in the other and sealed. The kit is very small and used for a single patient.

We consider that in laboratories working in smallpox areas the kit should be very similar to that of the WHO: a lancet or a Hagedorn needle in closed test tubes; 6 to 8 clean slides; tin foil or parchment paper for covering the slides; two small glass test tubes or vials that can be sealed; 6 to 8 capillary tubes in a test tube; two sterile swabs in test tubes; plasticine for sealing the capillary tubes; cotton wool, gauze compresses.

The whole should fit into a small metal box, using cotton wool, silica gel, calcium chloride, sawdust, etc. to keep the box dry.

This box is introduced into a second one likewise sealed and containing absorbant. The two boxes are then kept in a cardboard, wood or plastic box or in polyethylene bags.

When collecting, only the inner box is taken into the patient's room; the others are left outside to avoid contamination.

After collecting the pathologic products and putting them into the box the latter is wiped with a disinfectant; then once outside the room it is introduced into the second metal box. Together with the card accompanying the product, the box is put into polyethylene bag or other packing.

The card is completed and the two boxes closed only after the person having performed the collection has carefully disinfected his hands.

Transport of the products is done with the utmost care so as to avoid all possible risk of spreading the infection.

Today, refrigeration is no longer considered necessary because the virus is sufficiently stable at room temperatures provided the transport is not delayed. When possible refrigeration should be employed, the blood samples being frozen after separation of the serum from the clot.

WHO specialists recommend sending the products by special messenger or by any speedy and safe means available (train, air, etc.). The parcel should be precisely addressed with a warning concerning the infectivity of its content and should only be opened in the laboratory to which it is addressed. When the laboratory is in another country the stipulations of the international conventions must be complied with.

3. 3. Laboratory tests in the diagnosis of smallpox

Some techniques, which will be described in detail, may be successfully used for establishing a diagnosis of smallpox even in the laboratories with less sophisticated equipment; among these are: microscopic examination of stained smears, precipitation in gel and isolation of the virus on the CAM of the embryonate chick egg, currently used in the laboratories of India. The WHO specialists consider these methods as *compulsory* for any laboratory that has to establish a diagnosis of smallpox.

To these may be added examination in the electron microscope, which during the last few years, especially in centres in which smallpox was imported into Europe, has been recommended as a very rapid and sure diagnostic technique.

Although immunofluorescence is applied to a lesser extent in smallpox cases it may be useful since it is a rapid method of diagnosis.

Where any case of smallpox becomes a problem of vast implications – in the intact countries – the infection has also to be confirmed by more delicate methods, among others is experimentation on animals, which can only be done in well equipped laboratories.

3. 3. 1. Microscopic examination of the stained smears

The smears, collected at the patient's beside under the conditions described, are used as such. From the dry material on the test tube wall a suspension is prepared with a small amount of liquid over a surface measuring 1 cm in diameter.

The argentic impregnation method, as practised by Indian investigators, differs very little from that recommended by the WHO

Materials

(i) fixation solutions containing:

acetic acid	1 ml
formaldehyde	2 ml
distilled water	100 ml

(ii) the mordant contains:

tannic acid	2 g
phenol	1 ml
distilled water	100 ml

(iii) The ammonia silver nitrate solution is prepared from a silver nitrate solution (5%) in distilled water and ammonia solution (10%) likewise in distilled water. To 180 ml of the silver nitrate solution ammonia solution (ca 20 ml) is added drop by drop, with continuous stirring until a weak opalescence appears. At the moment of use the ammonia silver nitrate solution is diluted: 1:10 with distilled water.

Staining technique

Staining is carried out as follows:

(1) The slide with the smear is introduced in the fixative or the fixative solution is added to the slide; fixation lasts 1 minute.

(2) The smear is washed with distilled water.

(3) The smear is covered with a phenol tannic acid solution and the slide heated over a flame until vapour arises (30–60 seconds).

(4) The slide with the smear is washed with distilled water.

(5) The smear is covered with an ammonia silver nitrate solution diluted 1:10 and again heated until vapour arises (in 30 seconds).

(6) The smear is washed with distilled water, dried and examined in the microscope with an immersion objective lens.

It is recommended to fix the preparation in Canada balsam and cover with a cover-slip, because cedar oil sometimes wipes out the smear.

The following is the Gispen method of staining, adopted by the WHO specialists.

Materials

(i) The fixation and mordant solutions are prepared as in the laboratories of India.

(ii) The ammonia silver nitrate solution is prepared with a silver nitrate solution (10%) in distilled water and ammonium hydrate solution (1%) (prepared each time) obtained from a mother solution (25%). The technique is the following:

In a flask well rinsed with distilled water a drop of ammonia solution (25%) is pipetted, then silver nitrate (0.5 ml 10%), after which distilled water (20 ml) is added. The ammonia solution (1%) is added drop by drop, dissolving the ammonia silver precipitate (as a rule 1.4—1.8 ml); then a drop of (10%) silver nitrate is added causing opalescence of the solution.

Staining technique

(1) Fixation (for 1 minute), washing with water and drying the smear.

(2) Introduction of the smear into xylol for 5 minutes, then drying.

(3) The smear is covered with phenol treated tannic acid, the slide is heated over a flame until vapour arises (about 1 minute), then,

(4) washed under tap water and rinsed in distilled water,

(5) covered with the ammonia silver nitrate solution, freshly prepared, and heated until vapour arises (about 2 minutes).

(6) The smear is washed and dried and examined in the microscope with an immersion objective.

Hence, the difference is only the preparation of the ammonia silver nitrate solution and the introduction into xylol for fixation.

Results

On the smears prepared from the early skin lesions (up to the vesicular stage included) characteristic punctiform elements can be seen, uniform in size and distribution.

The rounded viral particles have a diameter about three times smaller than that of the staphylococcus and stain brownish-black against a yellow background.

The positive result does not differentiate the vaccinia from the smallpox virus.

The smears prepared from the more advanced stages of pustulation and crusts contain, besides the viral particles, cellular fragments and associated germs; the interpretation of such smears is very difficult.

Examination of the stained smears gives the best results when the smears are spread very thin and when there are samples from several smears on the same slide.

On chickenpox smears the viral particles are round, poorly stained and at the limit of visibility; they can only be recognised in the early stages of the eruption.

A skilled, experienced examiner can readily differentiate between the two types of particles — smallpox or chickenpox.

The result should be considered only as likely positive and has to be confirmed at any rate by precipitation in gel and isolation of virus on CAM.

Before giving a negative result it is recommended to wait for the results of the other tests.

3. 3. 2. Examination of the smears by immunofluorescence

When the viral particle or antigen has been made evident, immunofluorescence will permit its identification owing to the specificity of the sera used.

In general, the results obtained by this method are in agreement with visualisation of the virus stained by argentic impregnation, but have the advantage of immunologic identification.

The results may be obtained within a few hours but as already stated, cannot be considered decisive in sporadic cases.

The short interval in which the result is obtained renders immunofluorescence of exceptional usefulness when screening of a large number of cases is necessary in a community in which smallpox is endemic or an epidemic has broken out.

For the diagnosis of smallpox, immunofluorescence may be used in two ways:

— by examination of the products collected from the patient;

— by examining the cell cultures or CAM inoculated with these products.

In the latter case immunofluorescence permits visualisation of the antigens and even of the viral particles before the characteristic skin lesions can be detected with the naked eye or by microscopy.

Materials and techniques

The *quality of the sera* used has a direct influence upon the efficiency of both immunofluorescence and precipitation in gel. The best product is the serum collected from smallpox convalescents; when not available, hyperimmune sera prepared by repeated inoculations of the vaccinia virus to rabbits and goats can be used.

In principle the rabbit hyperimmune serum is prepared as follows.

Initially, the rabbit is inoculated on the scarified skin with a rich amount of vaccinia virus; 2 to 4 weeks later, when the crusts form, the rabbits are again inoculated 2, 4 and 6 times at a week's interval, by intravenous or intramuscular route, with a vaccinia virus suspension containing 10^9PFU/ml^{-1}. Seven to 10 days after the last inoculation the rabbit is bled out.

In order to avoid the appearance of heterologous antibodies in the serum of the immunised animal, use is made at the inoculations of the vaccinia pulp collected from the same animal and if possible even from the site of the initial inoculation on the scarified skin.

Hyperimmune vaccinia antisera are used as such or in their globulin fractions; especially when labelled with fluorescein isocyanate gammaglobulin presents certain advantages.

Labelling of the antibodies with fluorescein isocyanate is done according to the classical methods.

The best *samples* for examination by immunofluorescence are those obtained from the early stage of the maculo-papules and vesicles, but the fluid aspirated from the pustules or crust suspensions can also be used.

If, however, we intend to use immunofluorescence from the beginning, then the centre of the maculo-papules must be scraped in order to collect a small drop of serum without any trace of blood; this droplet is collected on a slide. Several serum droplets from different lesions can be collected on the same slide.

In the stage of vesiculation or pustulation smears can be prepared with the aspired fluid. After raising the top of the vesicle or pustule the bottom is gently scraped with the tip of the bistoury or Hagedorn needle; the material obtained is smeared on a glass slide over an area of about 1 cm^2, without rubbing.

When the material has been brought in test tubes and has dried on the walls, as well as in the case of crusts, a suspension is prepared which is then smeared on a glass slide.

Although not a major indication, immunofluorescence may also be applied to the smears prepared from the patient's blood in the acute stage of viremia, or from the blood of contacts. In this case the smears are prepared directly, immediately after withdrawing the blood or later from the blood collected in test tubes; smears with red blood cells not included in the clot are preferable.

During the initial period, in the presence of lesions of the mucosa smears may likewise be prepared from the pharyngeal or conjunctival exudate collected by swabbing.

The smears are dried at room temperature and fixed in acetone for 10 minutes; fixation with dioxan should not be used as it destroys the viral antigen.

As fixation with acetone does not inactivate the smallpox virus, the slides have to be exposed to ultraviolet light, either before using acetone or afterwards (2 minutes exposure at a distance of 2 to 3 cm, up to 15 minutes exposure at 50 cm distance).

The smears are divided on the slides into two squares (drawn with a dermatograph or nail varnish) treated concomitantly, one with immune serum and the other with normal serum from the same species as the preceding one.

Staining may be done either by the direct or the indirect method.

For direct staining the smears are covered with conjugated immune serum on one of the squares and conjugate normal serum (from the same species as that in which the immune serum was prepared) on the other.

Staining lasts 30 minutes at room temperature, the slides being kept in a moist atmosphere (in Petri dishes with filter paper or cotton soaked with water).

For indirect staining, non-conjugated immune and normal serum are added to the two squares for 30 minutes.

The smears are then washed 2 to 3 times in a buffered solution: the excess washing fluid is absorbed with filter paper from the smears. Both smear surfaces are covered with conjugated serum containing antibodies to the globulin species that supplied the sera (immune and normal), used in the first phase.

Staining lasts 30 minutes, then, as in the direct method, the smears are washed three times in buffer and mounted in buffered saline solution, preferably gylcerinated, below a cover slip; after eliminating the excess fluid the edge of the cover slip is fixed with nail varnish.

In this way the preparation can be minutely examined and photographed without detaching the material from the slide.

Besides the control staining, performed on the same slide, staining will be done in the same way each time pathologic products are examined in smears prepared with smallpox or vaccinia virus suspensions. The activity of the sera will be controlled in the same way.

The working dilutions of the sera are usually indicated by the firm that supplies them; if not they have to be established in the laboratory.

For direct staining the smears prepared with vaccinia virus suspensions are covered with successive conjugated serum dilutions, using the highest dilution that offers net fluorescence.

With the indirect technique a series of smears prepared from the vaccinia suspension are covered with different dilutions of the immune serum and, respectively, normal serum; then to each series is added a dilution of the conjugate serum.

As a rule, the vaccinia antiserum prepared in the rabbit, diluted 1/10 and rabbit antiglobulin serum, prepared in the goat in a 1/5 dilution, insure convenient staining.

Results

In general, in smallpox, irregular areas with a bright, specific fluorescence can be recognised on the smears, as well as numerous points with a similar fluorescence that represent the viral particles.

(i) On the smears obtained from the imprints of the skin lesions the cells with viral antigen can only be discerned in some of the slides, but the number of particles is often sufficient for the diagnosis. Leucocytes may also be seen, some exhibiting an unspecific fluorescence.

(ii) The best results are obtained with the products obtained by scraping the bottoms of the skin lesions. The examination reveals isolated cells with a mass of specifically stained, intensely fluorescent antigen, and cells with large or smaller stained granules, likewise intensely fluorescent; extracellular small, specifically stained granules, the viral particles, can be seen. There are likewise frequent polymorphonuclear leucocytes some of which contain specific antigen; the leucocytes sometimes give an unspecific fluorescence that may be contrasted by staining with rhodamine conjugated albumin.

(iii) Good results are also obtained with the vesicular fluid, the fluorescent bodies being free in the fluid or at the periphery of the erythrocytes.

(iv) Sometimes in the acute phase of intense viremia, fluorescent bodies may be recognised at the edge of the erythrocytes in the blood smears.

In conclusion, the viral material appears in the form of elementary bodies or conglomerates, the latter especially within the cell.

On the smears of chickenpox or herpes lesions the immune sera corresponding to fluorescent elements can be differentiated by their granular aspect.

In some cases unspecific immunofluorescence is so marked that the test may be considered as useless. This generally occurs with older preparations kept at room temperature without being fixed; correctly collected and fixed products may be kept and even transmitted long distances for examination.

Notwithstanding, the fresh preparations may also present accentuated unspecific fluorescence, especially when the material has been obtained from pustules or crusts; therefore immunofluorescence is not specifically indicated in the eruption stages.

Immunofluorescence is particularly valuable not only for a rapid diagnosis in first cases but even for screening suspect cases during the outbreak of an epidemic. Today, this method offers new diagnostic possibilities in smallpox, especially when used for isolating the virus in cell cultures or the CAM. It is worthy of note that fluorescent antibodies permit visualisation and identification of smallpox antigen and viral particles in HeLa cells within the first 24 hours after inoculation, before the cytopathic effect appears and even before haemadsorption becomes positive. Balş and Căruntu (1965), studying vaccinia infection by immunofluorescence in various types of cell cultures, identified the antigen in HeLa cells after 6 hours and in human fibroblasts or KB cells after 8 hours.

3. 3. 3. Examination in the electron microscope

Examination under the electron microscope calls for a special equipment and personnel; under certain conditions results may be obtained within a very short interval even if the samples are transported to laboratories in other localities.

Therefore, we consider it useful to give certain indications concerning identification of the smallpox virus in the electron microscope.

Initially, it was given only perfunctory attention even by Rake *et al.* (1948), who were among the first investigators to use this method, but today its practical importance has been emphasised by an increasing number of research workers.

It is considered that osmic acid fixation and metal shading techniques do not give satisfactory results, whereas negative staining with phosphotungstic acid, permits a simple, rapid and sure diagnosis.

Which specimens can be examined in the electron microscope?

The *specimens* collected from the patients are sent to the laboratory under one of the following forms:

(i) *Dry material* on slides (smears) or the walls of test tubes: the material scraped from the macules or papules or the bottom of the vesiculo-pustular lesions, the vesicular fluid or pustular content.

(ii) *Fluid material*, in capillary tubes or test tubes: the vesicular fluid or pustular content, as such or diluted in broth, buffered saline solution, etc.

(iii) *Crusts, the walls of vesicles or pustules,* transported in flasks or test tubes.

Processing of material for examination in the electron microscope

For this examination the material must be suspended, using as little liquid for the dilution as possible.

The suspensions may be used as such or, as recommended by Lépine (1964), they are left to sediment for 10 minutes in very fine tubes, the supernatant liquid then being drawn off.

A drop of suspension is put on the grids previously prepared with Formvar or collodion.

When the droplet is almost dry (drying can be done in vacuum) the grid is washed thrice with distilled water. Washing consists in holding the grid with a very fine forceps and depositing a drop of distilled water on it, then eliminating the excess water by holding a piece of filter paper along its edge.

Impregnation with phosphotungstic acid is likewise done by putting the grid held with the forceps in contact with a drop of reagent consisting of phosphotungstic acid (2%) adjusted to pH 7 with potassium hydroxide (1%); Lépine recommends pH 7.2 and considers that the addition of saccharose (1%) facilitates impregnation.

The excess phosphotungstic acid is removed with filter paper after which the grid is left to dry.

Cruickshank (1966) *et al.* expose both surfaces of the grid to ultraviolet light in order to inactivate all virus particles. Lépine (1964) as well as Kempe and Vincent (1964), likewise sterilise the grid by exposing it to ultraviolet light as soon as the suspension droplet dries, but before staining.

Another variant of the phosphotungstic acid impregnation method, applied especially when fluid extracted from the vesicles is used, consists in mixing a droplet of fluid with one of phosphotungstic acid solutions in a Petri dish. After stirring by repeated aspiration with a pipette, droplets of this mixture are put on the grids; the excess is immediately aspirated. The mixture is very viscous and leaves a fine layer on the grid; sometimes the viscosity is excessive and the mixture must be diluted with the same phosphotungstic acid solution.

Examination is done at a magnification 20 000 up to 40 000 diameters, and the structure of the suspect particles is examined at even greater magnifications. Photographs of the more representative images may be taken. The results are sometimes obtained only after developing the film and examining it with a magnifying glass, until sufficient particles with a characteristic structure are identified.

A specimen is considered negative only after it has been examined for at least 10 minutes and has to be checked by examining a second sample from a different part of the product before giving a definite answer.

When the case is positive the results are conclusive in at most 10 minutes but the examination may last only 1 minute.

The technique recommended by Cruickshank *et al.* (1966) may be carried out in approximately 5 minutes after receiving the specimen.

Negative staining is superior to all the other methods (fixation with osmic acid, shading, etc.) both for the ease with which it is performed and for the certainty of identifying the smallpox virus.

Results

When the material is obtained from recent eruptive elements and its preparation is correctly done, the viral elements appear homogeneous and typical. Lépine (1964) who studied their morphology and measured them concluded:

(i) smallpox virus has the classical brick-like structure with rounded corners. The viral particles are almost cubic and, as a rule, isolated, measuring 200/250 up to 200/290 nm;

(ii) the vaccinia virus resembles the smallpox virus but is smaller, less symmetrical and is usually found in clusters;

(iii) the chickenpox virus has almost spherical elementary bodies, of various sizes, appearing in clusters in small numbers; however; the diameter of the elements is 160 to 200 nm, most of them measuring 180 nm in diameter.

According to Lépine, examination in the electron microscope permits differentiation of the smallpox virus from chickenpox virus but the difference between the former and vaccinia virus is more difficult to establish and implies appraisal of a statistical order.

Forming part of the same poxvirus group is the milker's nodule virus (BX virus, Orf), which is ellipsoid in shape, resembling a silkworm cocoon, with interpenetrating capsomeres.

The chicken pox virus is much smaller. Within an irregular envelope the virion has a round or hexagonal cross section, with empty capsomeres covering its surface, and closely resembles herpes virus.

In the vesicular fluid the particles can be readily seen, as in smallpox, and most of them are enclosed in an envelope.

In chickenpox, with extracts prepared from crusts the results are not so easily obtained because the viral particles are scarce, and, as a rule, they lack the enclosing envelope which holds their fine structure together; therefore before giving the result a careful prolonged examination is necessary, and even microphotographic control.

3. 3. 4. Detection of the antigen by precipitation in gel

Precipitation in gel used in the diagnosis of smallpox is adapted from Ouchterlony's double diffusion technique; a precipitation band occurs along the line where the antigen comes in contact with the antibodies.

The test may be performed with material collected from the vesicles, pustules or crust stages, provided there is a sufficient quantity that can be used in concentrated form; this is afterwards diluted in at least ten times its volume.

Necessary materials

1. For the precipitation in gel reaction, the material collected and dried on slides, fluids brought to the laboratory as such, or crusts can be used.

The antigen is prepared from the *dry product* on slides, by scraping it off 3 or 4 slides into a small volume of fluid not exceeding 0.2 ml. In practice this is done by adding about 0.2 ml of a buffer saline solution (which may also contain antibiotics as the suspension can be used for the inoculation of embryonate eggs), scraping the slide with the tip of the pipette and mixing it with the fluid. The suspension is then taken up again and added to another slide, which is scraped, the material suspended and taken up again, etc. until a sufficient amount of suspension is obtained for the test.

The dry specimen can also be scraped off the slides into a Berzelius glass and taken up with an adequate amount of fluid.

The fluid collected from the vesicles or pustules is used as such or diluted stightly by adding a buffer saline solution (0.1 ml), keeping enough suspension for the inoculation of CAM.

Of the complex procedures necessary in diagnosing a case of smallpox, precipitation in gel plays a major part when crusts alone are available.

The crusts represent the best source of antigen for this reaction. They are treated with buffered normal saline in very small amounts: 0.2 ml for 2 to 5 crusts, or 0.4 ml for 6 to 10 crusts.

Good results can also be obtained with a single crust in 0.1 ml but this extract cannot be diluted without lowering the sensitivity of the reaction; in general it is recommended that several crusts should be used chosing the "fullest".

The suspension is then left for 1 hour at +4°C for extraction of the antigen. If the suspension is too thick it is left to sediment or it is mildly centrifuged.

As recommended by Dumbell and Nizamuddin (1959), to avoid risk of contamination, the extract is treated with formaldehyde (0.2%); infectivity disappears within a few hours without diminishing diffusion of antigen in the gel and precipitation. Even if the requirements are for an immediate test, formaldehyde insures subsequent inocuity of the product.

The formaldehyde treated material can be kept several weeks for use, if positive, as a control in other tests. The extract for precipitation in gel is treated with formaldehyde only when we also have the necessary material for isolation of the smallpox virus.

The detection of antigen in the *blood* by precipitation in gel may be done both in the pre-eruptive prodromal stage and at the beginning of the eruptive period.

The antigen is present in the blood only in the severe cases so that the test may also indicate the prognosis.

As a rule the blood serum in which the viral antigen has been diffused is used but a suspension may also be prepared by trituration of the clot, or by lysis of the red blood cells in distilled water.

Positive controls are always used in the tests, namely:

(i) extracts prepared from the crusts of a patient and proved to be positive, prepared as shown and kept in the refrigerator;

(ii) extracts prepared from the CAM infected with variola virus; when this is used a further negative control sample must be introduced in the test, i. e. an extract prepared according to the same method from a non-infected CAM;

(iii) vaccine virus may also be used as positive control in different forms: antismallpox vaccine as such, extract of the CAM infected with vaccinia virus, extract of the material collected from man or from animals inoculated by cutaneous route with the vaccine virus.

2. *Immune sera* used in the precipitation in the gel test must have a rich content in antibodies. As for immunofluorescence the best sera are those collected from smallpox convalescents.

The sera obtained from persons vaccinated against smallpox do not have a satisfactory titre; not even the sera collected from freshly vaccinated subjects give good results. Antivaccinal sera in high titres are obtained by hyperimmunisation, which is difficult in humans. Hence, it is recommended to use hyperimmune vaccinal antiserum prepared in animals.

Current use is made of rabbits and goats for preparing the sera; the mode of preparation in rabbits has been described in the paragraphs on immunofluorescence. In any case, the sera should be prepared in special laboratories with good equipment, and means of control and storage.

3. The *reaction* is performed in gel in a concentration of 0.8—1%, using a purified gelose (as a rule Special Agar-Noble Difco) or agarose, pH 7 to 7.2 using either phosphate buffer or Prince buffer. The gel may be prepared in several ways.

For small amounts the following materials are necessary:

(i) 10 ml 0.15 M phosphate buffer, pH 7.2;
(ii) 10 ml 9% NaCl solution;
(iii) 0.2 ml 1% methiolate solution;
(iv) 0.2 g agar.

The mixture is boiled in a water bath until the agar is completely melted, after which the slides are prepared.

When larger amounts of gel are necessary then proceed as follows:

(a) Prince's buffer is prepared from the following stock solutions:

2 M NaCl (116.88 g l^{-1})	50 ml
0.5 M TRIS hydrochloric (60.57 g l^{-1}, pH 7)	20 ml
0.5 M EDTA (acid) (146.12 g l^{-1})	2 ml
2% sodium azide in distilled water	5 ml
Distilled water	adjust to 1000 ml

Both the stock solution and the working solution prepared according to the above formula are kept at +4°C.

The volumes of the specified stock solutions are prepared in limited amounts in proportion to the composition of Prince's buffer.

The 0.9% agarose gel is prepared by dissolving 4.5 g in 500 ml Prince's buffer in a 1 000 ml flask, in a water bath. It is stirred constantly while boiling, which is continued for 20 minutes after the agarose has melted. The 0.9% agarose is distributed into several flasks in order to avoid degradation of the whole stock by repeated heating.

(b) Large amounts may likewise be prepared in the same way with phosphate buffer according to the following formula:

NaCl	7.65 g
Na_2HPO_4	1.27 g
$Na_2H\ PO_4$	0.1 g
KH_2PO_4	0.2 g
Distilled water	adjust to 1000 ml

When Special Agar Noble or agarose is not available then gelose can be used, after previous purification. Gelose is prepared according to Lépine (1964), as follows:

A 2% gelose solution in distilled water is filtered in the open autoclave and poured into test tubes in 10 ml amounts

When required for use the content, of each tube is melted and mixed with 10 ml of a veronal buffer solution adjusting the final pH to 7.2. The gelose should be kept in sealed tubes to avoid alteration of its concentration.

Gelose fibre may be purified by washing under running water for several days, after which it is dissolved in hot water or NaCl solution. After gelatification it is cut into cubes and dialysed against distilled water for 2 to 3 days, then dried at 80°C and kept in closed jars.

Merthyolate (1:10 000) treatment is necessary to avoid contamination of the gelose.

4. For the reaction gelose is spread in thin layers on glass slides.

The slides must be free of scratches; in general it is preferable to use new slides.

When all the slides have been prepared, the gelose is dissolved by boiling, and maintained at a temperature above 50°C. It is then pipetted on to the slides, each slide receiving about 2.5 to 3 ml, which is spread evenly over its surface to give a layer 1 to 2 mm thick, which corresponds to 0.2 ml/cm^2. Care should be taken to avoid the gelose running over the edge of the slide.

The following procedure makes it possible to obtain a gelose layer of even thickness:

Two slides are separated by two rectangular pieces of glass (cut from another slide) and fixed together with rubber bands or clamps. The melted gelose is poured into the space between the slides; after it becomes solidified (in 10—15 minutes at room temperature) the top slide is slipped off the bottom one without damaging the gelose layer.

Holes are punched in the gelose layer, either all at once with a special tool or one by one using a plastic template placed above the slide. After punching the holes the gelose is removed with a sharp instrument (syringe needle, etc.) or aspired with a Pasteur pipette from the wells which measure as a rule 4 mm in diameter and are punched at a distance of 1 to 2 mm from each other. The pattern differs; some prefer, as recommended by Lépine, a central well with six holes around it, others have a pattern with only four wells as recommended by the WHO.

Thus prepared, the slides are kept in a moist atmosphere (Petri dishes with wet cotton wool); they must, however, be used within a few hours.

Gelose may also be obtained ready made on slides or in polyethylene boxes, with the wells already punched out, supplied in airtight containers so as not to dry.

5. The reagents are introduced into each well with a separate, finely tapered Pasteur pipette so as to fill the holes without overflowing on the gelose surface.

When using Lépine's pattern, the antigen is introduced into the central well and the different sera in the peripheral ones; for negative control of the reaction both the specific smallpox and vaccinia antisera can be used and the chickenpox and herpes antisera.

If the pattern with four wells is used (numbered 1 to 4) the reagents are introduced as follows:

(i) normal serum well 3;

(ii) antivaccinal serum well 1;

(iii) positive control antigen well 4;

(iv) suspect suspension well 2.

When several samples are to be examined the reagents are introduced in turn, into the corresponding wells using a single pipette for each; finally, in well 2 the suspect suspension is introduced using this time a separate pipette for each sample.

After filling the wells the slides are put into Petri dishes on adequate supports (glass rods, matches, etc.) and maintained in a moist atmosphere at room temperature until precipitation lines appear.

The positive slides may be kept after being dried in a thermostat at 37°C, for 20 to 24 hours and then stained.

Staining is usually done with a 1% amido-schwartz 10B solution in:

99% methylic alcohol	50 parts
distilled water	40 parts
acetic acid	10 parts

The slides are stained for 1 to 2 minutes, then washed several times with a solvent and dried at room temperature.

With this technique the precipitation lines stain blue. The slides may also be stained with 1% azocarmine, in which case the precipitation bands are red.

Results

The results may be obtained within a few hours, sometimes two but generally in 4 to 5 hours; however, in some cases the reaction only becomes positive after a longer interval.

The slides are read against a dark background, with a strong light and using a magnifying glass.

The slide may be read when the vaccinal antiserum (well 1) and positive antigen (well 4) form a precipitation line.

When the smallpox antigen is present in the gelose then precipitation lines appear between wells 1 and 2. The sample is considered positive when these lines join those formed between wells 1 and 4, a process that occurs subsequently (after 4 hours or more), indicating the antigenic identity of the suspect sample and positive control sample.

When precipitation lines appear between wells 2 and 3 (normal serum) the result is doubtful and demands new tests.

A negative result (precipitation lines only between wells 1 and 4) can only be considered as such after 24 hours. The test will have to be repeated if a halo surrounds the wells.

It is recommended using the smallpox and vaccinia antiserum in two dilutions since it has been observed that in the presence of a serum diluted 1:2 the weak reactive reactions (due to a smaller amount of antigen in the extract prepared) are more readily observed (Dumbell and Nizamuddin, 1959).

A precipitation line between wells 2 and 4 reveals the presence of antibodies in the sample examined and indicates that the patient is in a late stage of the smallpox or that he is suffering from generalised vaccinia.

Precipitation in gel may however give a negative result even when the patient has smallpox if the material is not correctly prepared; therefore, it should be recalled that a diagnosis of smallpox can only be excluded if inoculation on the CAM also remains negative.

3. 3. 5. Detection of the antigen by counter-immunoelectrophoresis

Counter-immunoelectrophoresis is now being used in various laboratories for the identification and titration of certain microbial (bacterial, viral) and parasitic antigens and the corresponding antibodies. It offers the advantage of a shorter reaction time and a greater sensitivity than precipitation in gel, previously described. The technique is based upon the property of endo-osmotic migration of gammaglobulins towards the cathode and of the antigen towards the anode when precipitation in gel is performed in an electric field. The antigen and antibodies placed in wells close to one another migrate in the course of electrophoresis towards each other, forming precipitation bands along the line where they meet.

In principle this technique is used under the same conditions as precipitation in gel.

Materials

(1) For counter-immunoelectrophoresis the same pathologic products are used, processing of the samples being similar to that for precipitation in gel.

(2) The immune sera are prepared as shown above.

(3) The reaction is performed in gel (agar, agar-agarose, agarose, 0.8—1%), prepared in barbital buffer at pH 7.8—8.2.

(a) Although in most instances counter-immunoelectrophoresis is performed in agarose gel, on the basis of our experience we recommend the use of a mixture, in equal parts, of Special Agar Noble and agarose, since the presence of agar favours the phenomena of endo-osmotic migration of gammaglobulins towards the cathode.

(b) The barbital buffer contains:

Sodium barbital	8.142 g
Sodium acetate $3H_2O$	6.476 g
Hydrochloric acid(1M)	14—16 ml
Distilled water	adjust to 1000 ml

The buffer reaction is adjusted with hydrochloric acid to pH 8.1—8.2, under control of the pH-meter. This buffer is used for the cells of the electrophoresis apparatus and for preparing the gel.

(c) As a rule it is recommended that the gel be prepared by dissolving Special Agar Noble (1 g) and agarose 1g in distilled water (200 ml). This is done in a water bath in a sufficiently large flask to permit constant stirring during boiling, which is continued for 10—20 minutes after the gel has dissolved. The solution is then distributed in flasks, in adequate amounts for the rate and volume of work required, in order to avoid degradation by repeatedly melting the gel.

In our laboratory we have adopted the following procedure:

A double concentrated gel is prepared in bidistilled water Special Agar Noble (2 g) and agarose (2 g) in water (200 ml) distributed into small flasks, which are then sealed. The barbital buffer is likewise prepared in a double concentration.

When necessary, equal amounts of gel and buffer are heated in a water bath and after the gel has melted, are mixed together. This makes it possible to prepare from the same gel and buffer stocks, a gel with a different ionic force. It has been recommended recently to use a discontinuous system (ionic force of the buffer in the gel at least half that in the cells) which may have certain advantages in terms of the antigen and especially of the available immune sera.

(d) For the reaction the microscope and projection slides (Kodak, 3.25/4 inches) are cleaned, all traces of fat are removed and the slides covered with an agar film (2% in distilled water). After drying, the agar-agarose gel melted in a water bath is added (2.5—3 ml for the microscope slides and 13 ml for the Kodak slides). After gelation the slides are kept in a moist chamber for 1—18 hours, during which the consistency of the gel increases; then the holes for the wells, 3 mm in diameter, are punched along two parallel lines 5—10 mm distant (centre to centre).

On the microscope slides three pairs of wells are obtained and on the Kodak slides up to eight wells. Three, instead of two rows, may be punched, a free space of about 10—15 mm then being left at both ends for applying the wicks.

(e) The immune serum is introduced in the wells towards the anode and the products to be examined and the control negative and positive antigen in the wells towards the cathode. These reactants are pipetted into the wells 5 to 10 minutes after starting the electrophoresis, the interval being necessary for balancing the gelelectrolyte system.

Electrophoresis of the reactants takes place in a rectifier cell system; the amount of buffer in the cells, the size of the binding wicks (1 to 3 layers of filter paper or Watmann paper No. 4) depend upon the cell type.

The migration current is adjusted to 1 to 3 mA per cm front (as a rule 8 mA for a microscope slide and 24 mA for a Kodak slide).

Results

The precipitation bands appear 30 to 60 minutes after the run is started, halfway between the antigen wells and the immune sera wells, but this also depends upon the ratio of one to the other.

When the pathologic product is poor in antigen (actually poor or consequent to dilution) premigration of the antigen may be started 10 to 15 minutes before adding the immune serum.

On reading a positive reaction, account must be kept of the control system in which the proportion of the reactants is chosen so as to insure optimal conditions for formation of the precipitate.

The results may be improved if the slides are kept after the reaction at + 4°C for several hours and after washing with phosphate buffer saline. Precipitation lines may not be visible immediately after electrophoresis, and washing with PBS wipes out false precipitation lines.

Counter-electrophoresis is more sensitive than precipitation in gel. Antigens with a titre of 1 : 4 by the latter technique are positive even when diluted 1 :16 with the former. Although more rapid and sensitive its specificity is similar to that of precipitation in gel in the variola vaccinia group, without, however, differentiating the antigens of these two viruses.

3. 3. 6. Detection of the antigen by complement fixation (CF)

Although not speciafically recommended, the variola antigen can also be detected by CF; its sensitivity, superior to that of precipitation in gel, permits detection of the antigen when there are very small amounts of material and particularly when blood is investigated at the beginning of the pre-eruptive stage.

If, with precipitation in gel, the antigen has to be concentrated and up to 12 crusts are necessary to obtain an efficient extract representing a 1 : 10 to 1 : 20 dilution, in CF a positive result may be obtained with a smallpox crust extract in a dilution of 1 : 3200.

The vaccine antiserum prepared according to this scheme is used in an optimal dilution established in the presence of a known antigen (chess-board titering).

Any of the CF techniques recommended in viral infections can be used; it is preferable, however, as recommended by Kempe and Vincent (1964), to perform the test in the cold (4 to 6 °C), overnight.

3. 3. 7. The isolation of smallpox virus

Smallpox virus may be isolated by several methods:

(i) Inoculation of sensitive animals (monkeys, rabbits, mice) by various routes (scarification of the skin or cornea, intratesticular or intracerebral inoculation) permits isolation of the virus and, moreover, offers the possibility of a differential diagnosis.

Almost exclusively used in the past, this method has been abandoned by most workers because it involes handling the infected animals, and the results are late and necessitate complementary identification tests.

Nevertheless, when the difficult problem of instituting measures that may influence the life of an entire community arises, the contribution of this method to the diagnosis, by identification of the virus — variola, vaccinia, varicella — must not be neglected.

(ii) Today, inoculation on the CAM of the chick embryonate egg is considered the most useful method as it is readily performed in any laboratory and the results are comparatively conclusive.

(iii) Inoculation in *cell cultures* is a sensitive, rapid method that can be applied in the laboratories using this technique.

3.3.7.1. *Isolation in the chorioallantoic membrane*

Inoculation on the CAM is recommended by WHO specialists for isolation of smallpox virus since:

(i) it is simple and readily performed;

(ii) it does not demand special equipment, material or training;

(iii) it gives a diagnostic indication within 48 hours (as a rule 72 hours) because chickenpox, herpes zoster and measles virus do not produce lesions in the chorioallantoic membrane, and those caused by smallpox, vaccinia or cowpox as well as of herpes, can be differentiated by their morphologic characteristics;

(iv) incubation of the inoculated eggs at different temperatures makes it possible to determine the type of variola virus isolated.

Materials

Smallpox virus is isolated in embryonate eggs incubated at 37 and 39 °C for 11 to 13 days, and when necessary, 10 or 14 day-old embryos can be used.

In the laboratories where this test is performed the eggs are put in the incubator twice a week (Thursday and Sunday or Wednesday and Saturday) so as to have 10 to 13 day-old chick embryos every day.

The laboratories to which the diagnosis of smallpox is assigned must have at least 12 embryonate eggs per week, increasing their number if an increase in the number of samples to be tested is foreseen, bearing in mind that 4 to 6 eggs are necessary for each sample.

A diagnostic activity implies an efficient supply of eggs (at least 80% fertilised), and correct management:

(i) the eggs must come from less than 2 year-old hens; they must not be exposed to the light or heat and must not be washed;

(ii) the eggs are kept a few days at a temperature of 10 to 15 °C before being put in the incubator;

(iii) the incubator is kept outside the laboratory so that the eggs may also be used for other purposes;

(iv) in the incubator, the necessary humidity must be maintained and the eggs must be frequently turned in order to permit a good development of the membranes, especially when the virus type will be determined on the basis of its thermoresistance. Nizamuddin and Dumbell (1961) recommend 12 days incubation at 38°C, turning the eggs every hour.

For *inoculation* of the CAM the classical method (Cajal, 1958) is recommended by the WHO specialists since a larger surface of the membrane can be inoculated and the lesions are more conspicuous.

The method consists of:

(i) viewing the eggs in the candler, choosing those with viable embryos, and marking the limits of the air sac and the centre of the CAM with fewer blood vessels;

(ii) the egg shell and its membrane are perforated over the air sac;

(iii) the shell is then cut over the CAM mark, avoiding injury to the shell membrane, or else a triangular piece is cut out (with a vial file, dental drill, etc.);

(iv) a drop of sterile saline solution is placed in the centre of the exposed shell membrane and introduced below the shell, almost parallel to it, with a sharp instrument (vaccinostylet, syringe needle); by raising the end of the stylet or needle in the hand the shell membrane is perforated and drops; the saline droplet is aspired. A new air sac is thus formed which can be checked in the candler. If necessary, the CAM can be dropped by suction with a rubber bulb over the air sac.

If the blood vessels remain stuck to the egg shell it means that the CAM has been perforated and the egg must be discarded.

In order to ensure maximum expansion of the new air sac inoculation is done before 30 minutes have passed and not later than 2 hours.

Inoculation is done with the syringe, introducing the needle about 5 mm in the orifice above the new air sac and depositing 0.1 ml suspension on the CAM. The egg is then inclined in all directions so as to spread the inoculum over the whole surface of the CAM.

The cut is sealed with a drop of petrolatum or a mixture of petrolatum-paraffin, or with a cover slip held in place by a rim of sterile petrolatum-paraffin, or with adhesive cellophane, etc.

The eggs are *incubated* with the inoculation hole upwards and, in most instances for isolation of smallpox virus, at a temperature of 35 to 37 °C, for 3 days; after 48 hours one egg of every sample is controlled.

Collection of the CAM can be done after removal of the embryo in a Petri dish. The following procedure is generally applied:

(i) the egg is put on a suitable support, disinfected, and the shell and its membrane removed over the CAM;

(ii) the CAM is cut off from its insertion upon the shell, washed in buffered saline solution (BSS) and examined in a Petri dish with BSS, against a dark background.

If the CAM will not be subsequently used for preparing suspensions for inoculation, it will be introduced into a 10% formaldehyde solution and examined.

The membrane may be examined also *in situ* after removing the shell over the inoculated area. The number and aspect of the lesions are appraised directly, after which a portion of the inoculated membrane is cut off, washed in BSS or formaldehyde solution and examined on a flat surface, usually a Petri dish.

Pathologic products and their processing for inoculation on the CAM. Viruses may be successfully isolated from products collected in any stage of the eruption, from the first macule or papule up to the last crust; as already mentioned, it may also be isolated from the blood in the initial stages of the disease, even in the course of the pre-eruptive phase especially in the severe cases.

Viruses may also sometimes be cultivated from the pharyngeal or conjunctival mucosa, the samples being collected with sterile swabs that are kept as such or in sterile broth.

When several samples are available, preference is given to the fluid collected from vesicles or pustules and the suspension prepared from crusts, spreading the material on slides.

The *dry material* brought to the laboratory on slides or in test tubes is scraped with the tip of the pipette and suspended in a very small amount of BSS with antibiotics, after which the procedure is similar to that described for precipitation in gel.

From the *vesicular or pustular fuid* 1:10 and 1:1000 dilutions are prepared in BSS with antibiotics (0.1 ml fluid + 0.9 ml BSS, and 0.1 ml of this dilution is diluted in 9.9 ml BSS). If the amount of fluid is too small, it is initially diluted in 0.1 ml BS and to what remains after precipitation in gel 0.5 ml BSS with antibiotics is added; considering this dilution as 1:10, the 1:1000 dilution is prepared from it.

The *crusts* as well as the top of the vesicles and pustules are triturated in a small amount of BSS with antibiotics and from this suspension the two dilutions mentioned are obtained.

If the material is processed for precipitation in gel what remains is suspended in 0.5 ml BSS with antibiotics; considering this a 1:10 dilution, the 1:1000 dilution will be prepared from it.

The suspensions are left to sediment for 10 minutes and the clear supernatant is used for inoculation.

The *blood* is another material that can be used for isolation of virus. As emphasised by Lépine (1964) the richest material for inoculation is the leucocytic layer.

When the collected blood forms a clot both the serum and the clot can be used for inoculation. The latter is triturated and suspended in 1:10 up to 1:1000 distilled water; by haemolysis the viral particles on the surface of the red blood cells are released.

As the samples collected from the skin are contaminated it is recommended to add antibiotics to the BSS, in general 500 IU penicillin ml^{-1} and 500 μg streptomycin ml^{-1}, eventually also 100 ml^{-1} μg neomycin ml^{-1}. Lépine, who considers smallpox virus sensitive to penicillin recommends that apart from the streptomycin and neomycin added to the extract, 500 IU penicillin and 100 μg streptomycin should be inoculated in the yolk sac immediately before inoculation of the CAM.

When the blood is collected under sterile conditions no more antibiotics are added to the material to be inoculated.

Results of inoculation on the CAM

(1) *Cultivation of poxviruses on the CAM* determines the following changes:

Proliferative lesions develop on the CAM, isolated or confluent according to the wealth of the suspension in viral particles and interval of the observations. The lesions, currently known as colonies, pustules or *pocks*, appear as punctiform thickenings with different features in terms of the virus cultivated.

According to Fenner (1968), the colonies may be:

(i) white, opaque;
(ii) white, opaque with a central haemorrhagic ulceration
(iii) pale, with a central haemorrhagic ulceration;
(iv) ulcerohaemorrhagic lesions (in some cases only).

The vaccinia virus produces numerous, thick, large lesions (2 mm at 48 hours and 3 to 4 mm at 72 hours), whitish opaque and well developed over the surface of the membrane. The rest of the membrane exhibits intense congestion and eventually necrotic lesions. As a rule the lesions produced by vaccinia virus have puffy edges, are larger and flatter than the smallpox lesions, and show a tendency to confluence.

Smallpox virus produces characteristic white-opaque lesions that are small (a diameter of less than 1 mm at 48 hours), slightly bulging, with a net contour; subsequently the colonies develop and reach a diameter of 1 mm at 72 hours.

In general, cowpox virus produces haemorrhagic lesions when isolated, then generates small, whitish opaque mutants, slightly bulging similar to the variola lesions; Netter and Lapeyre (1964) showed, however, that in terms of the cowpox virus strain, lesions similar to those of smallpox can be obtained from the beginning.

The neurovaccine virus produces haemorrhagic lesions on isolation.

Although it is considered that the vaccine virus grows very well and frequently kills the chick embryo, the aspects observed depend upon very many factors so that they cannot be considered as characteristic. Moreover, it is often very difficult to differentiate between the lesions produced by the two viruses — variola and vaccinia.

Monkeypox virus forms small whitish lesions similar but not identical to those of smallpox virus.

Of the viruses that do not form part of the variola-vaccinia subgroup, *chickenpox virus* does not produce lesions of this type (colonies) on the CAM, only irritative lesions or small, uncharacteristic granulations, and *herpes virus* produces small white-opaque lesions, which may however sometimes become large and ulcerated, similar to the vaccinal ones.

●

In the case of high virus concentrations, the confluent lesions thicken the entire CAM and render interpretation very difficult. For this reason inoculation of the diluted product is recommended since isolated lesions can be obtained; the WHO recommends 1:10 and 1:1000 and Lépine 1:10 and 1:100 dilutions.

From each dilution 2 or 3 eggs are inoculated; one of them is opened at 48 hours and the others at 72 hours.

A diagnosis of smallpox must be based on several CAM lesions; one lesion is not sufficient. Hence it is necessary, as well as when uncharacteristic opacities appear, to inoculate a fragment of the membrane to other eggs.

●

Differentiation of variola virus from vaccinia virus, on the one hand, and of the other two types of variola virus, on the other, can be done by incubation of the inoculated embryonate eggs at different temperatures. It is known that variola virus develops better at lower temperatures than vaccinia virus. For isolation of smallpox virus the WHO specialists stress the fact that the temperature must remain below 37 °C (35 °—37 °C).

For the former differentiation the eggs are incubated both at 35 °—37 °C and at 39.5 °C. In the presence of vaccine virus, when incubation is done at 39.5 °C, the number of lesions for the same amount of product inoculated, is reduced by at most one logarithm; with cowpox virus the lesions are only slightly reduced. Even at high concentrations of the inoculated product, the smallpox virus is not able to produce a sufficient number of lesions at a temperature of 39.5 °C (Netter and Lapeyre).

Dumbell *et al.* (1961) apply the thermal tolerance test: 40.5 °C for vaccinia virus, 40 °C for cowpox virus, 39 °C for ectromelia virus, 38.5°C for smallpox virus and 38.25 °C for alastrim (variola minor). Therefore regulation of the incubator at 38.25 °C with variations of at most 0.25 °C will differentiate the latter two viruses: the alastrim virus no longer develops, whereas the virus of variola major produces lesions even if fewer and somewhat smaller.

It is, however, difficult, technically speaking, to control the incubator accurately. Heating with a continuous current of hot air has been recommended, the temperature being checked at least thrice a day, close to the eggs.

Helbert (1957) differentiates variola from alastrim virus by the mortality rate of the eggs inoculated on the CAM with different virus concentrations: at a concentration of 10^5 the smallpox virus kills all the embryos in 3 days, and alastrim virus only 60%; at a concentration of 10^3 the former kills 100% of the embryos in 6 days and alastrim virus only 10% in 7 days.

The differentiation tests are performed under the best conditions when the previously isolated virus is used and corresponding suspensions of the CAM are prepared.

3.3.7.2. *Isolation in cell cultures*

Cell cultures have offered the possibility of a complex study of the characteristics of the variola vaccinia subgroup, either studied directly or completed by further examinations: fluorescent and immunofluorescent microscopy, electron microscopy, serological tests.

In practice, cell cultures make it possible to establish a diagnosis of smallpox, to isolate and identify the virus and to differentiate it from the vaccinia virus.

Although isolation and identification of smallpox virus and different types of cell cultures have been demonstrated (Marennikova *et al.*, 1964; Lépine, 1964; Netter and Lapeyre, 1964; Benda *et al.*, 1968) they are not yet currently used for the diagnosis of smallpox.

Isolation of smallpox virus in cell cultures is done starting from the same material as that used for the inoculation of embryonate eggs.

The processing of the material follows the same pattern. The suspensions are diluted 1:10 or 1:1000; a too highly concentrated suspension in viral particles produces rapid degeneration of the cellular layer, without forming characteristic lesions.

Numerous cell types have been used for isolation of smallpox virus: the best results are obtained in primary cell cultures. Monkey kidney, human embryo kidney or bovine kidney cells are generally recommended, but newborn hamster or rabbit kidney cells, guinea-pig lung and chick embryo or human embryo fibroblasts are likewise very useful.

Similar good results have been obtained with some cell lines such as: RB (bovine kidney), HeLa, KB, HEp_2.

It should be emphasised that the aspect of the lesions produced by smallpox virus and also the other poxviruses of the vaccinia subgroup differ more or less in terms of the cell type used for the culture.

Worthy of note with the smallpox virus is the proliferative character of the lesions produced, forming true small tumours in the cellular monolayer. These proliferations may sometimes be taken for cell clusters not dissociated by trypsinisation. Subsequently, necrosis of the tumoural centre takes place, and the lesion extends and forms plaques.

In general, vaccinia virus produces lesions which have the aspect of irregular plaques from the beginning.

In both cases if the material contains little virus the development of the lesions is delayed.

Neutralisation of virus with immune sera helps to identify it, by protecting the cells.

The preparation and utilisation of cell cultures is done according to the current techniques described in the textbooks.

For isolation of the variola virus, the following must be kept in view in inoculating cell cultures:

(i) the cells to be inoculated have to be repeatedly washed after removing the growth medium that contains human or calf serum.

(ii) incubation is prolonged up to 3 hours after inoculation in order to insure retention of the smallpox virus which is adsorbed more slowly than other viruses;

(iii) to add to the inoculum a small amount of growth medium (preferably without serum) in order to favour adsorption and emphasis of the cytopathic effect;

(iv) the plaques are more readily observed if the maintenance medium contains calf serum in a concentration of 10% instead of 2.5%. Kempe and St. Vincent (1964) prefer, however, a maintenance medium without serum or with 2% chicken serum.

We shall present only the results that can be currently obtained in the laboratory.

Chick embryo fibroblast cultures

The smallpox virus does not produce plaques within the first 3 days of incubation at 37 °C; the characteristic colonies appear after this interval (proliferative lesions) and the delayed multiplication must be taken into account.

Vaccinia virus generates plaques with irregular edges, measuring 1 to 3 mm in diameter, which may be revealed by adding a 1% neutral red solution.

The lesions have small, rounded cells, sometimes with slightly detachable processes, and the plaques are surrounded by multinucleate cells or cell clusters.

The plaques produced by cowpox virus are somewhat smaller than those of the vaccinia virus. Varicella virus does not multiply and herpes virus does not as a rule produce plaques within 3 days.

Monkey kidney cell cultures

Variola virus produces lesions of the proliferative type mentioned, and after 3 days microscopy reveals irregular plaques similar to those of vaccinia and cowpox virus; along their edge rounded, ballooned or even giant cells can be seen.

Vaccinia virus generates lesions with the characteristic aspect described, measuring 1 to 2 mm in diameter.

The cowpox lesions are similar but smaller.

Herpes virus develops only slightly, as in cell lines, and the plaques gradually extend covering the entire cell layer with rounded cells.

On the other hand, varicella virus grows very well although it adapts with greater difficulty to this type of cell than does the variola or vaccine virus.

In *cell line cultures* (HEp_2, HeLa) the smallpox virus produces characteristic lesions after 1 to 3 days — tumourlike growths in the form of small whitish spots, which under the magnifying glass appear as small, refringent plateaux.

On adding 1% neutral red these variola tumours stain bright red, and the remaining cell pink, followed afterwards by central necrosis and the formation of irregular plaques that cannot be distinguished from those produced by vaccinia virus.

The vaccinal lesions appear in the form of a plaque; cowpox virus has a similar effect.

Herpes virus produces plaques with rounded cells which extend throughout the cellular layer but seldom produce giant cells; staining differentiates the typical nuclear inclusion.

Varicella virus does not grow in HEp_2 cell line.

●

It is not the lesions alone that help us to identify the virus type; haemadsorption of the chick red blood cells likewise help identification.

Similar to the vaccinia virus, the variola virus grown in cell cultures adsorbs the red blood cells at the level of the lesions; some cowpox virus strains do not produce adsorption.

When herpes and varicella virus multiply they do not produce haemadsorption.

More readily performed and more precise, immunofluorescence may be used for examining cell cultures, to reveal the antigens and viral particles in the cells. For this purpose it is recommended the cells should be cultivated on slides introduced into tubes (Barski tubes). The slides are removed from the tubes and fixed in acetone according to the conventional procedure. The virus is inactivated by exposure to ultraviolet light either before or after fixation.

Cells detached from the slides may also be used for preparing the smears.

The specific antigen is localised in the cytoplasm, taking the form of strongly fluorescent cytoplasmatic masses of antigen.

Examination with a powerful objective lens reveals the antigen as larger or smaller intracellular particles and sometimes as viral particles lying close to the damaged cells.

The smallpox virus may be visualised and identified by immunofluorescence within the first 24 hours, a period in which haemadsorption does not take place; nor does the cytopathic effect appear. When the products are poor in particles this is very important.

Incubation of the cell cultures inoculated with pathologic products at different temperatures (eventually completed by immunofluorescence examination) may differentiate the variola from the vaccinia virus. For this purpose HeLa cells are inoculated with different dilutions.

One series of test tubes are kept at 36.5° C and another series at 40° C; the slides are stained with fluorescent sera after 24 to 28 hours.

The variola virus is slower to multiply than vaccinia even at 36.5° C (fewer, smaller, more isolated foci) and at 40° C does not multiply at all, whereas vaccinia virus may be detected in isolated cells or in small foci containing a few cells.

Similarly, Moustardier (1973) recommended that the pathologic material should be inoculated in primary monkey kidney or rabbit kidney cell cultures and incubated at different temperatures, some at 36° C, others at 39.5 °C:

(i) in case of smallpox virus, only at 36° C and only in monkey kidney cells are small pseudotumoural formations with progressive central necrosis produced at 48 hours;

(ii) vaccinia virus develops well in both types of cells at both 36° C and 39.5° C, producing typical necrosis lesions;

(iii) varicella virus does not produce a cytopathic effect in these conditions.

In conclusion we recommend:

(i) when possible to use also cell cultures for detection of smallpox virus;

(ii) the cell cultures used for this purpose should be grown on slides whose staining, after incubation at different temperatures, will reveal the viral antigen and character of the lesions;

(iii) when the cultures are grown in test tubes without slides, examination of the cellular suspension by immunofluorescence permits visualisation and identification of the antigen, without differentiating between smallpox and vaccinia; this may be attempted by incubation of the test tubes at different temperatures and determination of the presence or absence of the two viruses by haemadsorption.

3.3.7.3. *Identification of the isolated viruses*

The isolated viruses, recognised by their multiplication characteristics may be further identified by various procedures.

Immunological procedures differentiate viruses of the variola-vaccinia subgroup from those with which they might be confused (chickenpox, herpes and herpes zoster).

Precipitation in gel, complement fixation and haemagglutination-inhibition (HAI) are performed *in vitro* in the presence of vaccine antiserum; suspensions prepared from the CAM or cell cultures in which virus developed are used as antigen.

The same suspensions, in the presence of the vaccine antiserum can be employed for neutralisation tests in cell cultures and on the CAM.

Inoculation of the CAM with different dilutions of the isolated virus permits recognition and reconfirmation of the character of the lesions produced.

Incubation of the eggs inoculated on the CAM with a virus suspension in different concentrations, at varied temperatures, permits their identification according to the schemes of Dumbell (1968) and Helbert (1957).

Finally, inoculation of the isolated virus to animals (monkeys, rabbits, mice) furnishes interesting data for its identification (Table 3.3).

TABLE 3. 3.

Procedures and results of inoculation to laboratory animals for the diagnosis of smallpox

Inoculated animal	Route of inoculation	Lesions * caused by		
		Variola virus	Vaccinia virus	Varicella virus
Monkey	Scarified skin	Maculo-papular eruption along the scarification lines	Reduced reaction of the vaccinal type	None
	Intratesticular inoculation	Voluminous orchitis	Slight or no orchitis	None
Rabbit	Scarified skin	Rare, mild eruption, without inflammatory reaction, cannot be passed in series	Intense papulo-pustular reaction, inflammatory reaction also maintained at subsequent passages	None
	Intratesticular inoculation	No apparent orchitis alterations	Orchitis	None
	Scarified cornea (Paul test)	Mild, or absent keratitis; small vesicles along scarification lines after 24—48 hours	Keratitis with suppuration and opacification of the cornea	None
Mouse	Intracerebral inoculation	None	Lethal encephalitis	

* Lesions produced 3 days after inoculation.

Of the techniques used, inoculation on the scarified skin of the rabbit is readily performed and differentiates the smallpox from the vaccinia virus after this has been previously recognised as forming part of the variola-vaccina subgroup.

Intradermal inoculation, or on the scarified skin, of a suspension prepared from the CAM on which the vaccine virus has developed (1:1000 dilution of a 20% suspension) produces necrotic lesions after 3 days, the pathogenicity of the virus being maintained at the following passages.

As a rule, smallpox virus does not generate lesions on the rabbit skin, not even at the first passage; the few strains that may however produce lesions at the first passage lose this property at the second passage.

Monkeypox virus resembles vaccinia virus in that it produces lesions on the scarified rabbit skin both at the first and the subsequent passages, but the lesions may be differentiated by their haemorrhagic and necrotic character.

3. 3. 8. Serologic methods of diagnosis

A significant increase in the antibody titre to viruses of the variola-vaccinia subgroup is of diagnostic importance, especially in the atypical or subclinical cases.

Detection of antibodies in patients who have not been vaccinated during the previous two years, as well as a fourfold increase in their titre in two successive serum samples collected at 4—6 days interval from more recently vaccinated subjects may be of great help.

The following tests are used for detecting the presence of antibodies:

(i) *Haemagglutination-inhibition (HAI)* which with a titre of at least 1:160 in the serum collected between the third and fifth day of the disease from a non-recently vaccinated person, as well as a fourfold increase of the titre in paired samples, suggests a diagnosis of smallpox.

(ii) *Complement fixation (CF)* in non-vaccinated persons only reveals the presence of antibodies after the tenth day in a titre of 1:16 or more; in previously vaccinated subjects this titre may be determined starting from the sixth day. A single sample collected in the acute phase permits a presumptive diagnosis if the titre is sufficiently high and the patient has not been vaccinated. A fourfold increase of the titre in paired samples permits a definite diagnosis.

(iii) The *neutralisation* test is carried out in cell cultures or embryonate eggs; as antigens, a vaccinia virus or variola virus suspension is used (especially when inoculating CAM as the variola virus produces smaller colonies). When cell cultures are used the serum titre is established in terms of the dilution that permits a 50% survival of the amount of virus.

On titering in the CAM the amount of antibodies that reduces by 50% the number of lesions induced by the virus suspension (as a rule 50—100 colonies) is determined.

Titres of 1:8 or 1:16 eventually appear after primary vaccination but are much greater in the case of smallpox (1:64 up to 1:1024).

(iv) *Precipitation* in gel, applied according to the usual techniques, reveals precipitating antibodies in the sera of patients or convalescents, but the results are negative in the sera of vaccinated persons (Moustardier, 1973).

3. 3. 9. Organisation of a diagnostic activity

The laboratory is obliged to confirm or invalidate a diagnosis of smallpox within as short an interval as possible: to confirm in order to justify and to extend the measures of control — to invalidate in order to avoid measures that might deeply upset the entire activity of the affected community.

Subsequently, when the existence of a focus of smallpox has been established, the laboratory has the task of a correct and rapid screening of the suspects.

This screening is necessary in order to avoid:

(i) overcrowding of the smallpox hospitals with patients who afterwards will be found to be suffering from another disease;

(ii) quarantine for persons whose diagnosis was invalidated but who cannot take up their activity because they have come in contact with the virus.

Both in the former case and during an epidemic outbreak the role of the laboratory is to establish the diagnosis in conditions that may be confused with smallpox owing to similar characteristics:

(i) eruptive diseases in which the skin lesions develop passing through the successive stages of maculation, papulation and vesiculation, category that includes herpes, herpes zoster and especially chickenpox;

(ii) atypical postvaccinal reactions with a pathologic character, presenting a localised or generalised rash in persons that have received the smallpox vaccine and are within an epidemic focus.

Moreover, account has also to be taken of the cases of smallpox that may evolve atypically, particularly in persons with a partial immune status.

In general, any person with fever coming from smallpox areas is considered suspect and investigated; this approach is also adopted towards persons who have come in contact with a subject coming from contaminated areas.

A presumptive diagnosis may be established within the first 24 hours after receiving the pathologic samples. When the smallpox focus is already known, light microscopy and precipitation in gel may be applied for a rapid diagnosis. In India these methods are considered as basic. When possible, to these may be added examination by immunofluorescence and by electron microscopy.

Parallel to these investigations which may sometimes be performed within 2 to 6 hours, several tests confirming the initial diagnosis within 2 to 4 days and considered as semi-rapid, should be carried out:

(1) Inoculation on the CAM of embryonate eggs. From each product using two dilutions, two series of 4 to 6 eggs are inoculated; one series is incubated at 36.5 °C and the other at 39 to 39.5 °C. Examination of the CAM 3 days after inoculation will give indications concerning the virus. In order to determine the smallpox virus type a third series of inoculated eggs will be incubated at 38.25 °C.

Apart from the morphologic characters of the lesions on the CAM, the virus can also be identified serologically (HAI, CF or immunofluorescence), by neutralisation or by inoculation on the scarified rabbit skin.

(2) Inoculation in two dilutions will also be done in cell cultures (HeLa cells or chick embryo fibroblasts), incubating one series at 36.5 °C and another at 40 °C. The cell cultures will be examined at 24 to 48 hours by immunofluorescence.

In a subsequent stage, if necessary, and if not performed in a previous stage, the smallpox virus isolated is typed, by its cultivation on the CAM in successive dilutions and by its maintenance at a precise temperature; the results are obtained in 5 to 10 days so that it may be considered a slow test.

The various degree of complexity of the tests also differentiates the level at which they can be performed:

(i) light microscopy, immunofluorescence, precipitation in gel, isolation of virus in embryonate eggs are accessible to any laboratory with minimum equipment;

(ii) electron microscopy and differentiation of the viruses by incubation at different temperatures in embryonate eggs or cell cultures, as well as inoculation of animals demand special equipment and facilities and personnel with great experience; such tests should be performed in specially prepared centres.

3. 3. 10. Organisation of a laboratory for the diagnosis of smallpox

Included in the plans for controlling smallpox there must be laboratories equipped for carrying out tests necessary to facilitate speedy and accurate diagnosis. In the event of an epidemic, such laboratories, together with trained personnel, should be anticipated and organised accordingly.

The laboratory should be in a central position so that infected material collected from suspected cases of smallpox quarantined in hospitals assigned for the purpose, will not be delayed in transit. If necessary, in the event of a sudden outbreak, an emergency laboratory within a particular area may be set up and staffed with personnel from nearby hospitals or other medical sources. Wherever it is situated, the primary concern must be to prevent the spread of virus and to protect personnel, which means that the building in which the laboratory is housed must be adequately staffed by trained people.

Ideally, the laboratory should be separated from any other activity, but WHO considers that a separate section in a laboratory fitted up for microbiology, with one or two rooms approached by means of a lobby, would suffice. The accommodation, however, would need to consist of a cubicle in which staff can undress, another in which they can don protective equipment, and a bathroom with showers.

The following is an outline of the requirements for such a laboratory as recommended by the WHO:

A poster, with SMALLPOX – ENTRANCE FORBIDDEN, must be put up at the entrance to the laboratory.

In the lobby, the personnel will take off their clothes and shoes and put on laboratory footwear.

Outside the door of the lobby and that leading to the laboratory there must be a bath mat dipped in disinfectant; the shoes worn in the laboratory must be disinfected at the door to the lobby on leaving, and the town shoes at the entrance door.

In the laboratory, next to the door are clothes hangers with the protection equipment.

In the laboratory it is preferable to have separate containers for the various operations (unpacking of the pathologic products sent, preparation of the suspensions necessary for precipitation in gel and inoculation, immunofluorescent staining and argentic impregnation, microscopy examination, inoculation and collection of the CAM, inoculation of cell cultures, examination and preparation of the eggs for inoculation, incubation of the eggs and inoculated cell cultures, etc.).

The laboratory should have about four tables (one of which should be a large one), cupboards for the reagents and glassware and laboratory stools.

The apparatus, instrumentation and glassware should be adequate for the unit to be able to carry on independently: incubator and thermostat for the inoculated eggs and cell cultures, refrigerator, microscope, candler, trays for the eggs, staining baths, stands for the test tubes, sterilisers, instruments, glassware, etc.

On each table there must be a vessel containing a disinfectant and glasses with bichromate solution. The table should be covered with gauze dipped in disinfectant.

In the same room, or an adjoining one, there should be an autoclave and a large boiler for sterilising the protection equipment.

Central laboratories should have several rooms (2 or 3) since their activities are complicated by identification tests (incubation at different temperatures, serologic tests, inoculation of animals, etc.). They may be part of the building that houses the section where smallpox vaccine is prepared.

The floors and walls of the laboratory must be constructed of a material that is readily washable with disinfectants and the working places arranged so that they can be sterilised by ultraviolet light. Similarly, the ventilation should be designed so as to avoid contamination of the atmosphere in the environment.

The laboratory personnel must be vaccinated against smallpox once a year; this also refers to all persons coming in contact with the laboratory or members of its staff, even if only once. It is also recommended that families of members of the staff should be vaccinated every year.

Personnel must wear adequate protection equipment: a gown closed at the back, trousers, shoes or sandals, close-fitting headgear, spectacles and protection mask and gloves.

Footwear is disinfected by rubbing them on the disinfected mats; the spectacles and gloves are disinfected at the end of the day's activity and the linen gowns, etc. sterilised by boiling daily.

Protection equipment is changed every day.

Personnel must be constantly reminded that working in the laboratory, whatever its nature, is extremely dangerous.

Parcels containing pathologic products sent for examination must be opened only in the laboratory. The packing and containers in which they came must be burnt immediately or, when this is not possible, dropped into a bin with disinfectant.

The boxes with the samples are wiped on the outside with a cotton pledget dipped in disinfectant and after removal of the samples they are put in the bin containing infectious material, to be autoclaved.

The slides with the smears, capillary tubes, Pasteur pipettes or tuberculin syringes are placed in Petri dishes after use and put in the bin containing infected material, as also are the vaccino-stylet or Hagedorn needle used for collecting the material.

During processing of the sample its infectivity should always be borne in mind.

If the smears are not exposed to ultraviolet light then the solutions with which they are treated for staining will be considered as contaminated and disinfected, as well as the staining bath which is autoclaved at the end of each day's work.

All the objects worked with (pipettes, test tubes, slides, capillary tubes for preparing the suspensions from the dry material on slides, from vesicular or pustular fluid, or crusts) are considered contaminated and put into a bichromate solution, in a bath with disinfectant, in the bin containing infected objects or in the instrument box in which they are sterilised.

For precipitation in gel, Dumbell and Nizamuddin (1959) recommend treatment of the extract with 2% formaldehyde for several hours. If this is not done, then the slides with gelose and the dishes in which the slides are kept must be regarded as highly infectious and handled as such.

Similar measures must be taken when inoculating embryonate eggs; the test tubes, syringes, pipettes, etc. will similarly be disinfected.

On opening the inoculated eggs the same precautions must be taken; a gauze strip dipped in disinfectant is laid on the table and after the operation with all the fragments of egg-shells, etc. that have fallen on it, it is placed in a bath with disinfectant. After removing the CAM the rest of the egg is put into a vessel with disinfectant.

The instruments used, one set for eggs inoculated with the same product, must be sterilised and the support upon which the operation takes place will be put into the container with disinfectant.

Nothing must leave the laboratory unless it has been completely disinfected.

Registers or other documents must be exposed to the sun or ultraviolet light, and each file for at least two hours first on one side and then on the other.

For destruction of smallpox virus we recommend:

(i) immersion in boiling water, with a very rapid effect;

(ii) immersion overnight in a disinfectant (2.5% lysol, quaternary ammonium compounds in concentrations of at least 1%, 5% chloramine, 2.5—5% formaldehyde).

3.3.10.1. *Laboratory diagnosis of suspect cases outside known foci*

From the suspects isolated in hospital or in quarantine in a special unit, the product which indicates the stage of the disease is collected according to the previously mentioned instructions by the head of the staff working in the smallpox unit and sent to the laboratory.

The samples are packed and transported by the hospital sanitary personnel to one of the specialised centres indicated in the previously agreed plan.

The samples must reach the laboratory within a few hours (2 to 4) of being collected. The centre to which they are sent must be notified beforehand by telephone so that all should be ready for carrying out the tests as soon as they arrive.

When possible, a skilled technician from one of the specialised centres should go to the unit where the case has been admitted, bringing with him the necessary material for the tests.

These will be performed in a special room in which no other products are being tested, the material thus being processed under conditions of complete isolation.

For the tests that cannot be carried out under these conditions, the processed and inactivated products (for electron microscopy) or not inactivated (for isolation of the virus) are forwarded to the specially equipped institutes.

The presumptive result is the direct responsibility of the worker from the specialised laboratory and will be confirmed by this laboratory.

3.3.10.2. *Laboratory diagnosis within a declared smallpox focus*

In the hospital assigned to the isolation of smallpox patients a special laboratory is opened in which complex tests may be performed. It should be equipped with:

(1) Direct microscopy;
microscopy by immunofluorescence;
precipitation in gel;
counter-immunoelectrophoresis.

(2) Eventually, isolation and identification of virus in embryonate eggs. In this unit all the products to be sent to the specially equipped laboratories for detailed examination will be prepared (electron microscopy, isolation in cell cultures, identification of the virus strain isolated).

The work in this laboratory is done by a skilled team, previously instructed according to the operative programme, at district level.

The laboratory must be completely isolated either in the hospital building or in the special unit for smallpox. In the former, it should not communicate with the rest of the hospital.

The laboratory prepares the kits for sterile collection of the products and hands them over to the smallpox unit.

The products for the laboratory tests must be collected by the doctor looking after the patients, who must be instructed in this connection. The laboratory personnel does not come in direct contact with patients but only with their products.

If all measures of protection are complied with in the laboratory the members of the laboratory staff need not be kept in quarantine; this personnel may eventually have to move to other sites if the epidemiologic situation demands it.

●

In conclusion, the preparations for a smallpox laboratory consist of:

(1) The organisation of special laboratories within existing institutions: assigning and organising the rooms, establishing the

circulation of the personnel and pathologic products, drawing up the rules of protection, etc.;

(i) basic training of the personnel, eventually in laboratories abroad where a diagnosis of smallpox is currently performed;

(ii) to draw up the documentary material and instructions for the laboratory teams working in the field;

(iii) periodical refresher courses with the laboratory personnel working in the field, using simulated situations and adequate products collected from laboratory animals (monkeys, rabbits);

(iv) centralised preparation in the specialised laboratories of the materials for diagnosis (immune sera for precipitation in gel, fluorescent sera, media for precipitation in gel, the necessary reagents and substances) and their distribution to the field laboratories;

(v) preparation of the kits for collecting the products and their improvement.

(2) Selection of the laboratory teams that will work in the smallpox foci;

(i) periodical immunisation against smallpox of the personnel delegated for work in these laboratories;

(ii) designation of the rooms that will serve for the laboratories concomitant with that of the units, or hospitals assigned to the care of smallpox patients and quarantine of suspects;

(iii) equipment of the laboratories with the necessary utilities (collection kit, light microscope and with fluorescence, incubators, the current laboratory glassware and equipment, protection clothes, etc.);

(iv) to establish the items that will be given to other sanitary units and prepare the necessary stocks;

(v) to provide for means of transport.

4. Epidemiology

4. 1. Epidemiologic factors

4. 1. 1. Source of infection

1. Today it is considered that man is the only natural host of smallpox virus; no data exist proving that any other animal species may be infected in nature. Confirmation of this fact is of the greatest practical importance for eradicating the disease.

The role that monkeys might play in the maintenance and spread of variola has been the subject of numerous investigations and discussions.

Several natural infections similar to smallpox have been reported in non-human primates, of which three were in our times; in the absence of virological studies it cannot be asserted that they were caused by smallpox or another poxvirus. The last episode, occurred in Djakarta in 1949, when two orangutans contracted the disease in the course of an epidemic; the isolated virus was considered to be of human origin (Arita and Henderson, 1968).

Various monkey species have been infected by airborne route with variola virus, but positive results have seldom been obtained and the disease has presented features rarely encountered in humans (shorter incubation, less intense exanthema, lasting for a shorter interval and localised especially over the flexion areas, low mortality).

In the course of serial passages of smallpox to *Macacus irus* by aerosols the virus progressively loses its pathogenicity, with spontaneous interruption of the transmission; other monkey species in South America and Africa, which usually come in contact with man, have rarely proved receptive to smallpox infection. These data suggest that no simian reservoir of smallpox virus exists (Noble and Rich, 1969).

Similarly, eradication of the disease in regions with large monkey populations (Panama, Philippines) make it very improbable that these countries should be reservoirs of smallpox virus, moreover, no cases of smallpox among the monkeys of India, a country with a high endemicity, have been reported.

In 1959, the *monkeypox virus* was recognised as a specific virus. It has the characteristic properties of the vaccinia subgroup (brick-like shape, measuring 200/250 nm, produces cytoplasmic inclusions, causes skin lesions and gives cross reactions in serologic tests); some of its properties are closer to those of the human variola virus (small colonies on the CAM and characteristic lesions on the rabbit cornea), others to those of vaccinia virus (can be carried through serial passages on the scarified rabbit skin, kills three-week old mice following intracutaneous inoculation, forms plaques on chick embryo fibroblast cultures). In contrast to the viruses mentioned, the monkeypox virus produces haemorrhagic lesions in the rabbit.

In *Macacus irus* and *Macacus mulatta* species monkeypox appears to be more severe than smallpox (Arita and Henderson, 1968).

Until now monkeypox virus has only been isolated from monkeys in captivity in the course of epizootic episodes (10 in the last 15 years) recorded in the research institutes of U.S.A., Denmark, France, the Netherlands, as well as from human cases. The virus has not been isolated either from wild monkeys or from other mammal species that live in human smallpox areas, and which might eventually represent the true virus reservoir.

Serologic tests did not reveal significant neutralising antibody titres to poxviruses in monkeys in Asia and Africa (of some 14 species investigated), and the presence of such antibodies in other mammals of the same regions do not permit any conclusion, if account is kept of the similar antigenic structure of numerous pox viruses.

Therefore nothing can as yet be asserted concerning simian monkeypox focuses. It has been suggested that monkeys, as with man, are only accidentally contaminated with this virus and that the true monkeypox virus reservoir is one of the inferior mammals.

Man does not appear to have any special predisposition for the monkeypox virus; when several focuses developed among monkeys in captivity there was no human case; on the other hand after the first case was diagnosed in 1970, a few (17) human cases appeared naturally in Zaïre, Liberia, Nigeria, Sierra Leone and Ivory Coast.

Clinically, although the patients — mostly children and non-vaccinated persons — presented with classical symptoms of smallpox, in some a limited rash with a rapid evolution developed: four deaths were recorded.

Epidemiologically, the cases appeared in regions remote from one another, in areas free of smallpox. The source of infection could not be detected for any of the patients and no monkeys were known to become infected in the same areas.

Although it was emphasised that as a rule the infection is not transmitted to non-vaccinated persons in close contact with the patients, cases of transmission from human to human have recently been described.

The virus strains isolated from patients have the same properties as those isolated from monkeys in captivity (Table 4.1).

TABLE 4. 1.

Properties of monkeypox virus and its relationship to smallpox virus *

Samples	Monkeypox virus (Copenhagen)	Viruses isolated in Africa	Smallpox virus
Pustules with a haemorrhagic centre on CAM	+	+	—
Develops on the rabbit skin	+++	+++	—
Forms plaques in cell cultures	large	large	small or hyperplastic foci
Grow at 39 °C on CAM	+++	+++	—
Pathogenicity for mice and chick embryos	+++	+++	+
Produces haemagglutination	+++	+++	+

* after WHO Chronicle 1971, *25* (8), 398.

In this connection, it is worthy of note that three poxviruses, differing from monkeypox virus but very similar to smallpox virus were isolated from the kidneys of apparently healthy monkeys; the relationships with these viruses are not yet clear.

As also mentioned at the conference held in March 1969 at Moscow, the WHO Expert Committee on Smallpox Eradication (1972) considers it very little likely that an animal smallpox reservoir does exist. However, the possibility that certain non-human primate species may be infected with smallpox virus, on the one hand, and, on the other, the presence of monkeypox virus able to determine the clinical picture of smallpox in man, demand that further studies should be carried out in this direction.

At any rate, as underlined by the investigators who met in Geneva in 1973, monkeypox virus does not appear to represent a danger to the programme designed to eradicate smallpox (Henderson and Arita, 1973).

2. If man represents the only *source of infection* then this refers in general to the *smallpox patient* considered as infectious from the *onset of the disease* and *until the last crust has fallen.* The maximum period of contagiousness is the 3rd to the 4th day after the onset of fever (therefore with the beginning of the eruption) and up to the ninth day.

However, recent findings (Sarkar *et al.*, 1973 b) have drawn attention to the fact that family contacts of smallpox cases may eliminate variola virus from the throat within the first 4 to 8 days after a case has appeared in the family and that of these, some excreta contacts developed the disease subsequently. The proportion of excreta contacts is at any rate greater in the non-vaccinated subjects than in those with a past history of vaccination.

When investigations are carried out in order to detect all the contacts of a smallpox patient one must bear in mind that the infective period begins in the incubation phase. Similarly, account has to be taken of the fact that the contacts may also spread the disease and must be treated accordingly.

In the case of variola major, the virus strains have the same high degree of infectivity; in variola minor the epidemiogenic potential is less.

3. The largest amount of virus is eliminated by the respiratory route, especially as the oropharyngeal enanthema is intense. Patients with accentuated inflammation of the upper respiratory tract eliminate particularly large amounts of virus (for instance the *princeps* case in the 1970 focus in West Germany).

Sarkar *et al.* (1973 b) showed that patients with the haemorrhagic and confluent forms eliminate more virus through the pharynx than those with mild forms of the disease, and the periods of excretion are longer than in the latter group.

Smallpox virus is also eliminated through skin lesions. From the beginning of the eruption there is a large amount of virus in the deep layers of the epiderm; the lack of permeability of this tissue prevents it, however, from passing exteriorly, but this occurs subsequently with rupture of the vesicles and pustules (in the mouth and respiratory tract early erosion of the mucous lesions produces massive early contamination of the saliva and secretions).

Hence, in the first stage of the disease infection is spread through the mouth and nose and afterwards through the skin lesions. The patient thus spreads the disease around him through his saliva, mucous secretions, vesicular and pustular contents and crusts contaminating all the objects around him. The virus is to be found on the patient's skin, linen, clothes and bedclothes, on the various objects he uses, on the floor, in the air, in the droplets eliminated in the dust raised from the floor after the droplets have dried.

The smallpox virus is also eliminated by the conjunctival mucosa when inflammation is present; from the cases without conjunctivitis Sarkar *et al.* (1973 a) a were not able to isolate the virus.

Smallpox virus is likewise eliminated in the urine in a high proportion of the patients.

In the haemorrhagic and confluent forms the titre of the virus found in the conjunctival secretion and urine is higher than in the milder forms of disease.

4. Variola virus has a strong resistance to the environment.

At Madras, Rao (1966) found virus in a closed room 45 days after the patient had left, and in hospital in New Delhi under similar conditions, virus was isolated after one year.

The virus survives longer the lower is the relative humidity.

In crusts, the virus remains alive after several months: at room temperature Mac Callum and Mc Donald (1957) established the presence of virus after 18 months and at +4° C after 10 years.

Wolf and Croon (1968) proved that alastrim virus can be found in the dry crusts after 13 years at laboratory temperature, with a marked decrease in titre but sustained virulence.

This high resistance in the environment was considered to be at the origin of some epidemics in England (1908—1952) when no source of infection could be found and imported cotton was considered to be the source of infection.

4. 1. 2. Transmission of the infection

1. Investigations carried out to date prove that almost all cases of smallpox infection are produced by direct contact with patients, especially after the eruption appears.

Variola has likewise been spread indirectly by the objects contaminated by the patient's secretions: linen which has accounted for contamination of the members of the family and of the members of the staff of a hospital or hotel.

The air in the vicinity of the patient is less infected by the droplets eliminated by coughing or speaking but rather by the dust particles rising from the floor or objects in the room upon which the droplets have dried.

Isolation of the virus from air samples taken close to the patient has proved difficult, the virus being in low concentrations and rapidly inactivated by the sunlight.

Airborne transmission is therefore limited to a certain distance from the patient but has been admitted for some cases appearing in hospital or even in its vicinity, in the absence of any direct contact with the patient.

Many investigators, however, consider that in such cases account has not been taken of contaminated objects (badly disinfected clothes, letters, etc.) nor were the exact movements of the person in question recorded. Rao (1966) showed that there are not more smallpox cases around the smallpox units than in the rest of the town (Madras).

Therefore, although variola is seldom transmitted from a distance, demanding in general a direct contact, dissemination of the disease may be airborne.

In the recent focus in West Germany (Meschede — 1970) infection was demonstrated to have been spread from the patient's room to the second floor of the building by air currents.

2. Variola may also be spread by accidental inoculation, by the cutaneous or blood stream route; this risk exists particularly for doctors, nurses and laboratory personnel. In such cases the incubation period is as a rule shorter. Before manifestation of the general signs a characteristic lesions appears at the site of inoculation between the seventh and eighth day.

3. The WHO findings do not confirm transmission of the disease by the ingestion of contaminated food. Nor are there any examples of infection or mechanical contamination by haematophage arthropods known. It has been demonstrated that flies can be contaminated by the skin lesions of patients but nothing indicates that there is transmission back again to man by this route.

Some animals, such as dogs, cats and birds, may be superficially contaminated or may swallow crusts but no evidence has been found of transmission to man by this route.

4. As regards the role of the smallpox patient in the transmission of the disease we must bear in mind the following facts:

(i) The very severe cases (early haemorrhagic forms) die before beginning to eliminate virus through the mucosa and skin.

(ii) Although patients with severe forms eliminate large amounts of virus due to their extensive mucous and skin lesions, they spread the infection to a lesser extent because they are more speedily admitted to hospital.

(iii) Patients with attenuated forms of smallpox have a lower infectious potential, eliminate less virus because they have fewer lesions, and recover within a shorter interval. However, such patients represent an important epidemiologic danger because they are not immediately hospitalised or are admitted to hospital with a mistaken diagnosis, and are not isolated, permitting spread of the disease.

(iv) The atypical forms and especially those without a rash but with enenthema (pharyngeal, influenza, pulmonary forms) have a high potential for spreading the disease because they are not recognised in time. These forms develop particularly in persons vaccinated in the past, so that variola is not even suspected. Even in the endemic regions the atypical forms are difficult to diagnose even by skilled physicians; they are all the more difficult in countries unfamiliar with smallpox, in which the general practitioners are unacquainted with smallpox clinics and still less with the atypical forms. In most of the smallpox outbreaks during the last few years in Europe (England, Poland, Sweden, Yugoslavia) the epidemic episode was generated by an imported, atypical or mild case of variola that was mistakenly diagnosed and the necessary anti-epidemic measures were not taken.

(v) The same form of the disease is more dangerous for those who have not been vaccinated; in vaccinated subjects the lesions are more discrete and less infectious, as is also proved by their crusts in which little virus is found.

(vi) In the case of haemorrhagic smallpox the virus from the skin appears to be less infectious than that from the blood (Rao, as well as Sarkar of Calcutta attributes this to the lack of antibodies in the blood of patients with the haemorrhagic form).

(vii) The majority of specialists agree that there are no healthy smallpox virus carriers.

5. As regards the transmission of the disease, it has been shown that the way in which contact is established between the patient and the healthy persons is of the greatest importance; the frequency and intimacy of the contact sharply influences the risk. Hence, in the transmission of smallpox the family may be considered to represent a fundamental epidemiologic unit. According to Rao (1966), 34.4% of the contacts in the same room as the patient fall ill; if they sleep in another room only 14.6% contract the disease.

The same statistics show that if the family have a single room to live in, 37.5% of the contacts fall ill as against only 17% in families with more rooms.

Most family contacts contract the disease in the course of the first week after the first case has been registered, which confirms the opinion that smallpox is transmitted particularly by the virus from the respiratory tract.

4. 1. 3. Receptiveness of the population

1. As a rule it is considered that any person who has come in contact with smallpox virus or that has not been vaccinated against smallpox is sensitive to the disease.

The receptivity of a population is modified both in the endemic regions, where a certain degree of immunity exists consequent to small, repeated infections produced by circulation of the virus and in the countries where smallpox vaccination has changed the immunological profile of the population.

Observations made in endemic regions show that not all those who are exposed contract the disease.

Studies carried out in India have shown that the non-vaccinated contacts contract the disease in a proportion of 37.5 to 45%; in Calcutta, in a farm with 100 non-vaccinated persons (for religious considerations), only 10 fell ill, and 9 died. In Africa, a similar study showed that of 36 persons exposed, 25 non-vaccinated family members contracted the disease. Hence, the proportion of cases ranges between 10 and 67%.

2. Individual receptivity is modified once the disease has passed; a second recurrence is extremely rare (1 in 1000 cases).

3. The spread of smallpox in a community is directly connected with the degree of protection of the population following vaccination.

The first vaccination confers an almost total resistance immediately after being done but disappears with time. Revaccination regularly performed maintains a high level of resistance.

The effectiveness of vaccination is demonstrated by the data recorded in India . In families living in a single room 45.1% of the non-vaccinated contacts fell ill and only 17.8% of the vaccinated ones.

The effect of the number of vaccinations can likewise be assessed by statistics from India which show that among the vaccinated and revaccinated contacts there was no case of disease, that among those who received only the first vaccination there were 2.8% cases and among the non-vaccinated 37.5%.

The interval elapsing after the last vaccination is reflected in the proportion of cases among family contacts: there was no case among those vaccinated within the last four years (average 1 year before), 0.7% among those vaccinated 5 to 14 years previously (average 7 years) and 1.5% among those vaccinated more than 15 years previously.

The eradication programme of WHO is based upon increase in the resistance of the population by vaccination of all age-groups and especially children. When the proportion of a community protected by vaccination (revaccination) is of at least 80% it is considered that the chances of spread of smallpox are practically nil in that community.

In the intact countries exposed to the import of the disease, a high level of immunity must be maintained by an adequate programme of vaccination.

4. The WHO experts do not consider that it can be proved that denutrition, the individual health status, or a concurrent infection influences the risk of smallpox infection.

4. 2. Measures of prevention and control

Of particular interest among the measures of prevention and control of smallpox are those taken as a rule in intact countries in the course of routine activity: immunisation against smallpox, and preparation of the material and personnel for the eventuality of variola cases.

The control measures refer to those taken by the sanitary authorities in intact countries when one or several cases appear.

4. 2. 1. Preventive measures

4.2.1.1. *Active immunisation*

Active immunisation is the most efficient means insuring protection against smallpox since it interferes at the level of the receptive population. The present programme for the eradication of smallpox throughout the world is based upon the administration of vaccine to an increasingly wider proportion of the population in endemic countries. Since variolisation is still used in some parts of the world we shall first describe it in the following paragraphs.

4.2.1.2. *Variolisation*

The oldest method of artificial immunisation, variolisation is still practised in some African countries, in some districts of Afghanistan, Nepal and Brazil but on a more and more reduced scale.

Today variolisation is done by inoculation of the variola virus by different techniques: incision or multiple punctures, scarification, fresh burns; other practices, including nasal inoculation, have been abandoned.

For variolisation the fluid or crusts of the skin lesions collected from mild cases or persons who were previously variolised are collected, dried, transformed into powders and sometimes mixed with herbs or oils.

Recent investigations on the crusts used for variolisation in Afghanistan have shown that they reach an infectivity concentration of $10^{6.2}$ to $10^{3.5}$ PFU g^{-1}.

Variolisation represents, however, a real danger since the virus used may trigger the true diseases and start infection within a given area.

In Mali, 18.3% of the variolised subjects contracted the disease, whereas in a neighbouring non-immunised community the proportion of cases was of only 12.5%. Variolisation, however, modified the mortality rate: in the latter group 11 of 65 patients died whereas in the former group none died.

Variolisation creates epidemiological difficulties and demands identification of the source of each new outbreak in the respective areas.

4.2.1.3. *Smallpox vaccination*

In the programme for the eradication of smallpox particular stress has been laid on the quality of the vaccine and the use of efficient vaccination techniques.

Smallpox vaccine is a suspension of vaccine virus prepared from the vaccinal pulp obtained by scraping the lesions produced by inoculation to young animals on the scarified skin. The smallpox vaccine is almost exclusively prepared today in calves.

It is worthy of note that in 1921, Levaditi and Nicolau (see Nicolau and Cajal, 1962) adapted a vaccinia virus to the rabbit brain and with the neuro-vaccinia thus obtained vaccinated more than two million people in Spain.

Smallpox vaccines have likewise been prepared in chick embryonated eggs and cell cultures; the latter have proved particularly efficient when tested on humans but are still the subject of experiments before being introduced into general practice.

Smallpox vaccine is delivered either as a fluid, glycerine representing the vaccine virus stabiliser, or lyophilised, which is a particularly stable form easily transported and stored, and therefore recommended in the tropical climate of the countries with endemic smallpox.

The WHO recommends the exclusive use of lyophilised vaccine for endemic countries or those exposed to imported cases; at present the fluid vaccine is hardly used at all in any of the countries with endemic smallpox. At the beginning of 1970, there were on the WHO registers 64 laboratories preparing the lyophilised vaccine in 54 countries; in Romania this vaccine is prepared by the Cantacuzino Institute, Bucharest.

A good vaccine must comply with certain norms concerning its potency and stability.

It must contain 10^8 PFU ml^{-1} (infectious plaque-forming units, determined by titrating on the chorioallantoic membrane of the embryonate egg), although at the first vaccination it was found that a weaker vaccine could also give a maximum proportion of 'catches'.

The proportion of vaccines checked by the WHO and found to meet with the norms was 74% in 1969 as against 43% in the first year of the eradication campaign (1967).

The stability norm demands that a lyophilised vaccine should contain 10^8 PFU ml^{-1} after being kept 28 days at 37 °C.

It is important to reduce bacterial contamination in the preparation of the smallpox vaccine; apart from the rigorous measures of sterility taken on preparing the vaccine, antibiotics (penicillin, neomycin) or phenol have been introduced in the vaccine.

Various methods of purification of the smallpox vaccine are now being studied; treatment of the vaccinal pulp with fibrocarbon gave higher take rates than non-purified vaccine in revaccination.

4.2.1.4. *Vaccination technique*

In general, the best site of inoculation is the outer aspect of the arm, above the insertion of the deltoid muscle or slightly behind the midline; this zone is readily accessible and the vaccinal lesion less exposed to injuries or dirt.

In India, the primary vaccination is done at this site and the second on the anterior aspect of the forearm; this facilitates the control and differentiation of the two cicatrices.

Inoculation is not preceded by disinfection of the skin; however, when necessary, one may wash the skin with soap and water, then rinse and dry it. Disinfection was found to inactivate the vaccine virus rather than kill the germs on the skin. When indicated, however, disinfection may be done with ether and the vaccine applied only after the skin is completely dry.

At present several methods of inoculation of smallpox vaccine are used: single or multiple scarification, multiple pressure, a rotary lancet, bifurcated needle or needleless jet injector; the WHO specialists consider that the last two methods give the best results.

(i) *Vaccination on the scarified skin* consists of a linear scratch done while applying the vaccine on the skin with an adequate instrument (vaccination lancet, syringe needle or Hagedorn needle, tip of the bistoury, etc.). The instrument should be sterilised by boiling or heating over a flame; at any rate it must be perfectly dry and cold in order to avoid dilution of the vaccine or inactivation of the vaccine virus.

With the instrument, a drop of vaccine is taken from the vial and put on the skin; then the skin is scratched, as a rule along a single 6 mm line; in some countries two neighbouring lines are scarified.

Scarification has to be sufficiently deep so that a slightly sanguinolent serosity appears after about 30 seconds. The vaccine may be pushed into the scratch with the tip of the instrument. The place is allowed to dry without wiping it or exposing it to the sun or heat. It is not necessary to apply a dressing but it is preferable to put on clean linen. Scarification continues to be the most frequently applied method owing to its simplicity, the possibility of interpreting the results correctly and its proven efficiency over the years.

(ii) *Vaccination by the multiple pressure method.* The same instruments are used: through the vaccine droplet about 30 scratches are made by pressing the instrument held tangential to the skin within an area, of at most, 6 mm in diameter. The pressure exercised must produce a pinkish serosity at the site of vaccination. This technique does not differ from the former as regards preparation of the site of inoculation, sterilisation of the instrument, post-vaccination care.

A similar principle is that of the ring procedure devised by Krawitz and applied in France by the Institut Mérieux. The ring is made of plastic and has several small sharp tips on which a small amount of vaccine is put. The ring is fixed on the tip of the forefinger, the protective cap is taken off and the ring with the vaccine is applied on the patient's skin.

(iii) *The rotary lancet* was tested in India; it resembles a double curette, with a five-toothed metallic disc attached to its extremity (1 cm in diameter). The teeth are applied onto the skin where the vaccine droplet is and then rotated once all round. The other spoon-like extremity is then used to introduce the vaccine in the scratches produced. The pre- and post-vaccinal preparations are the same.

(iv) *Vaccination with the bifurcated needle* has been widely applied in India in the course of experiments carried out by the WHO. The tip of the needle is dipped in the vaccine then the skin is pricked in at least 15 places over an area 6 mm in diameter. The skin must be perfectly dry so as not to dilute the vaccine and the amount must be very small. With this method very little vaccine is used, i.e. about 0.02 ml per person instead of 0.1 ml when scarification or the rotary disc method is used.

In India, comparative investigations were carried out on the two methods, bifurcated needle and rotary lancet, and the former proved superior. Children were vaccinated by the two methods one on each arm with three different vaccine concentrations (10^8 PFU ml^{-1}, 5.10^7 PFU ml^{-1} and 10^7 PFU ml^{-1}); irrespective of the vaccine concentration the proportion of major reactions was twice as great following vaccination with the bifurcated needle. Even in the children vaccinated with only one of the methods the results were in favour of the bifurcated needle.

(v) *Vaccination with the jet injector* gives very good results; it has been used experimentally in India but on a wide scale in vaccination campaigns in Africa (Millar *et al.*, 1969; Neff *et al.*, 1969; Roberto *et al.*, 1969).

The fluid is injected under pressure and penetrates into the skin without a needle. Jet vaccination realises an intradermal rather than a percutaneous penetration.

Numerous reports have shown that this jet vaccination is successful in 90 to 100% of the cases; vaccination performed in India by this method likewise confirmed these results.

Studies in which vaccination by jet injector and multiple pressure were compared showed that jet vaccination gives very good results with a concentration ten times smaller than that used in percutaneous vaccination (scarification, multiple pressure, etc.); similarly, testing in certain situations (primary vaccination, revaccination at long intervals after the first) showed that small doses are sufficient by this method. Therefore, jet vaccination through the skin is far more efficient and much less vaccine is lost than with the other methods.

With jet vaccination antibodies appear even in the absence of a local reaction; however, we must again insist on the fact that the relationship between serologic changes and the degree of protection have not yet been clearly defined.

The method presents numerous advantages:

(i) it permits a high efficiency rate – with the injectors about 700 to 1000 persons can be vaccinated per hour;

(ii) the risk of transmitting certain infectious diseases, among which is tetanus, is avoided (in Dakar where tetanus is endemic, in 1961 there were 32 cases of tetanus with eight deaths after vaccination, and in 1967 and 1968 after vaccinating 280 000 persons by jet injector there was no tetanus case);

(iii) there were no more complications than with the usual vaccination methods; febrile reactions did not exceed 5%.

The disadvantage of the method is that the entire amount of vaccine ejected is so very small (0.1 ml), that it does not always penetrate through the skin; however, this depends on the injector and can therefore be remedied.

The WHO specialists and those of India consider that this method is by far the best for the vaccination campaigns against smallpox. However, in spite of its numerous advantages the high cost of the injectors, their defects and the numerous problems raised by their upkeep and repair, as well as the fact that they bring together a great number of people, render the method with the bifurcated needle more useful. The rate of vaccination by this method is fairly high (500 to 1000 persons per day, performing 100 to 200 vaccinations with the same needle).

Finally, mention should be made of the attempts at preparing a smallpox vaccine mixed with lanoline and fixed upon a plaster that is applied to the skin.

4.2.1.5. *The results of smallpox vaccination* (primary vaccination and revaccination)

Vaccination produces artificially in man an infection with local and general clinical manifestations that permit appraisal of the result concerning change in the immunologic structure of the organism and the induced degree of resistance to smallpox.

The manifestations of vaccinal infection following smallpox vaccination are characteristic of primary vaccination; revaccination may produce manifestations differing from those encountered after the primary vaccination owing to the modified immunologic reactivity of the body after its first contact with the vaccine virus.

The disease as manifested after smallpox vaccination is not a process met with in nature; it can at most be transmitted accidentally by the person vaccinated to the people he comes in contact with, but this is considered rather as a complication of vaccination.

The vaccine virus must be considered as a laboratory strain deriving from cowpox after repeated passages in man and animals; the natural infection of cows (cowpox) may be transmitted to man, the manifestations being similar to those of vaccinia proper, but representing a separate entity. Moreover, the viruses differ by their morphologic and cultivation characteristics.

Careful interpretation of the local manifestations of vaccinal infection is necessary for assessing the efficiency of vaccination both in the respective individual and at community level.

According to the indications of the WHO a vaccinal lesion should be interpreted as follows:

The *primary vaccination* is considered successful when it evolves according to a typical pattern. Between the third and fifth day a small papule appears at the site of vaccination and is rapidly transformed into a vesicle. The vesicle grows and is surrounded by a zone of erythema of various size; subsequently the centre of the vesicle falls in and the vesicular fluid becomes purulent.

During this period axillary adenitis and fever may develop. As a rule the pustule reaches its maximum size between the eighth and tenth day, after which a scab forms and falls off between the 14th and 21st day, leaving a characteristic scar. It is recommended to perform a control examination at the end of the first week after vaccination when the pustule or vesicle has formed.

This reaction is called the *primary* or *major reaction*, and implies a successful vaccination.

Immunity against smallpox begins approximately on the tenth day after the primary vaccination.

Revaccination is considered successful when multiplication of the vaccine virus takes place, manifested as the control performed after 6 to 8 days by a pustular or vesicular lesion, or by a clearly outlined area of induration or congestion surrounding a central lesion that may be an eschar or an ulceration. This type of reaction is qualified as a *primary* or *major revaccination reaction.*

Another type is the *accelerated* or *vaccinoid reaction* which generally occurs in subjects with a residual immunity resulting from a previous immunisation (due to the disease or to vaccination). The reaction is maximum between the third and seventh day after inoculation. The local lesion is smaller than after the first inoculation. These two types of reaction, *the primary or major one and the accelerated or vaccinoid one* are considered as confirming the success of the vaccination or revaccination. The Indian specialists insist upon the presence of the central vesicle as an obligatory criterion confirming immunisation.

Any other types of reaction after the first vaccination or revaccination are considered by the specialists of India and WHO as *equivocal.*

This category also includes the *immediate or early reaction* which is assumed to indicate sensitivity to the virus; it may appear both in immune subjects and in those sensitive to smallpox. The local lesion is as a rule papular, rarely vesicular; it develops on the second or third day after inoculation and rapidly disappears.

Interpretation of this type of reaction as a success or indication of immunity is a mistake, since experience has shown that it may result in the onset of smallpox cases, some lethal. The correct behaviour in such situations is revaccination.

Another mistake, sometimes followed by the appearance of smallpox cases, is the ease with which one accepts the idea that the absence of a postvaccinal reaction is equal to immunity. The absence of a reaction should be considered as the consequence of a technical mistake or a reduced immunogenic value of the vaccine or else that the respective person needs a more potent vaccinal lymph.

Several cases, some fatal, have been recorded in recent smallpox episodes in Europe and Asia in which the absence of a local reaction to revaccination was qualified as a state of immunity.

When vaccination is performed correctly and products of adequate potency are used the vaccine will take in 100% of the cases at the primary vaccination and 50 to 70% at revaccination in terms of the degree of immunity of the individual following earlier immunisations.

When the vaccinal reaction is considered as negative or equivocal, vaccination has to be repeated, eventually with a more potent vaccine.

In the Spanish army, Rodriguez and Santos (1965) after the first vaccine administration obtained 45 448 positive reactions in 67 170 vaccinated recruits. After the second vaccination another 2918 were positive, and after the third 1086, hence the proportion of successful revaccinations was 73.7%. Positive results can be obtained in over 90% of cases by repeating the vaccination.

•

Antibodies develop in the blood of the vaccinated and can be titrated by various tests: neutralisation, complement fixation, precipitation and haemagglutination inhibition. As neutralisation measures the anti-infectious potential of the individual it appears to be the best serologic immunity test.

Neutralising antibodies are detectable 10 days after vaccination as a rule in titers of 1:8—1:16 and 6 days after revaccination in much higher titres. Persistence of the neutralising antibodies exceeds that of CF precipitation and HAI antibodies.

CF antibodies are less often revealed after vaccination and revaccination.

Precipitating antibodies appear 10 days after the primary vaccination, on the seventh day after revaccination, and persist for about a month.

HAI antibodies may be detected in the blood after the tenth day, reaching a peak concentration after 3 to 4 weeks then gradually disappearing; after revaccination HAI antibodies may not be present.

As regards the correlation between the presence of antibodies and the state of immunity there is no general consensus, but it may be assumed to indicate an acquired resistence without being able to ascertain to what extent. The absence of antibodies reveals nothing at all. Hence the result of vaccination is generally appraised according to the local response.

During the last few years numerous attempts have been made to correlate the evolutive type of the cutaneous lesion with the level of serum antibodies stimulated by vaccination, in order to establish their protective level. Some authors consider the serologic test more important and reliable than the skin response, but it is very difficult to apply in the mass of the population. Sarkar argues that whenever the pustule forms antibodies are always produced.

In this connection there was a question of whether there is a serological difference between the vaccinates with postvaccinal encephalitis and those without. It was hoped to find a criterion of diagnosis in postvaccinal encephalitis and other clinical syndromes that complicate smallpox vaccination; no essential serological difference was found in the subjects with nervous complications, nevertheless, higher precipitating antibody titres were found in the vaccinates without complications.

A general agreement exists concerning the passage of antibodies through the placenta: between the third and sixth months after birth, antibodies in a protective titre are to be found in about half of the children and in only one-third towards the end of the first year. This is an argument lending support to those who sustain that vaccination should be performed in the first year of the child's life when the mother's antibodies may neutralise the circulating virus and prevent nervous complications.

Antibodies do not pass through the milk or do so only in a very small proportion.

Sometimes the maternal antibodies concentrate in the fœtal blood in higher titres than in the mother's.

Many investigations have been carried out regarding vaccination and the relations between mother and child. It is considered that the vaccinated child can be breastfed by the mother with smallpox and, inversely, the child with smallpox can be fed by the vaccinated mother.

4.2.1.6. *The complications of smallpox vaccination*

Although today the vaccines prepared are well purified and controlled complications still occur.

The WHO experts consider that these complications may be classed in two large groups:

(i) with rare sequelae or death;

(ii) with permanent sequelae and death

The incidence of complications is higher after the primary vaccination than after revaccination, and is closely linked to the age at which the first vaccination was performed: the older the vaccinates the higher is the frequency of complications.

The complications of smallpox vaccination have been studied in detail. Two large groups may be differentiated: those *affecting the skin, and complications involving the central nervous system.* In addition there are some rare complications due to localisation of the vaccine virus in various organs.

Cutaneous complications. This group includes several categories with a different mechanism of production and practical importance:

secondary bacterial infection, autoinoculation of the vaccine virus in other regions of the body, and generalisation of vaccinia in various clinical forms. In the great majority of cases skin complications appear in the vaccinates but may sometimes also appear by direct contact in other persons.

Secondary bacterial infection is rarely encountered today and readily treated with adequate antibiotics.

The vaccine virus may be seeded by scratching in various areas of the skin causing typical single or multiple lesions, seldom appearing successively. Repeated autoinoculation in an individual can only occur up to the ninth day after vaccination since immunity develops within this interval.

In persons with a predisposition to eczema the issue of auto- or hetero-inoculation may be very severe, especially at the level of the eyelid (a lesion similar to a stye, sometimes with fistulisation and enlargement of the preauricular lymph nodes) or cornea (keratitis of various types, sometimes followed by ulceration of the cornea and loss of vision, especially when complicated with vaccinal uveitis and retinopathy, or suprainfected with bacteria, Johnes and Huseine, 1963).

The most important complication is known by the name of generalised vaccinia and was recognised from the very beginning of smallpox vaccination.

Clinically, generalised vaccinia appears in three forms, of which *gangrenous (progressive) vaccinia* and *eczema vaccinatum* present a high proportion of deaths. The WHO lists these forms among the severe complications. Generalised vaccinia may likewise take on a mild, benign character.

Eczema vaccinatum is extremely severe and occurs as a rule in persons suffering in the present or past from eczematous manifestations, a situation in which vaccination is formally contra-indicated. A haematogenic rash develops at the site of vaccination after 5 to 11 days, then in the skin areas previously affected by eczema, and finally extends to the neighbouring healthy areas.

The WHO Committee emphasises the importance of this form in contacts with vaccinates; about 50% of the cases of *eczema vaccinatum* are produced by heteroinoculation of persons with latent or active eczema coming in contact with a vaccinated individual. The lesions frequently become infected with bacteria resistant to antibotics (as a rule staphylococci).

Less frequently seen, but of exceptional gravity, is the second form of generalised vaccinia — *progressive (gangrenous) vaccinia* — which occurs in one to two cases per million of vaccinates. Although the lesion at the site of inoculation evolves at first normally, it does not heal and necrosis develops and extends from area to area until a large zone is affected. Due to haematogenic generalisation similar lesions appear on other parts of the body and become necrotic; metastases are encountered at other sites on the skin, bones or organs. The disease lasts for weeks and even months, and death usually occurs from toxiinfection with septicaemia.

Gangrenous vaccinia appears to be caused by an immune defect of the body, manifested by hypo- or gammaglobulinaemia and the absence of plasma cells (O'Connell *et al.*, 1964).

Neurologic complications. The frequency and gravity of the neurological complications have led several specialists to oppose routine smallpox vaccination in the countries from which the disease has been eradicated. They consider that the risk to which the vaccinates are exposed is not counterbalanced by the utility of vaccination.

Studies carried out to date have established certain particularities of these complications without, however, fully explaining their pathogenesi.

A problem not yet solved is that of the varied incidence of *post-vaccinal encephalomyelitis* (PVE).

Striking is the fact that the incidence of PVE differs from one country to another. Regional differences may also appear within the same country, or the incidence changes from year to year, ranging between 1:4000 and 1:100 000. The importance of age at the first vaccination has already been stressed.

As regards the causes that interfere in encephalitic complications we are no more advanced than we were 40 years ago (Herrlich, 1952; Baylet *et al.*, 1963). Of the numerous suppositions made we shall recall the following:

(i) intervention of the vaccina virus;

(ii) the possible reactivation of latent viruses;

(iii) when the former suppositions were not confirmed the encephalitic complications were considered as neuroallergical reactions;

(iv) a particular genetic character, proper to the vaccinated population has also been assumed.

PVE usually appears 8 to 15 days after vaccination of a non-immune subject (primary vaccination or revaccination after a longer interval in which the immunity induced by vaccination has disappeared).

The onset is sudden, with fever, headache, vomiting, loss of consciousness, sometimes with meningeal signs and paralysis. Convulsions and somnolence are gradually followed by coma and death.

In the cerebrospinal fluid there is an increase in the amount of proteins and cells, especially lymphocytes.

The EEG shows slow, high voltage waves, generalised in both hemispheres.

The histologic examination demonstrates perivenous microglial proliferation in the white matter of the cerebral hemispheres and brain stem, with extensive perivenous demyelination processes.

In contrast to PVE the vast heterogenous group of *postvaccinal encephalopathies* include various clinical pictures that appear between the second and 25th day after vaccination, more frequently around the eighth day. The onset is likewise marked by fever and convulsions, occasionally with hemiplegia and aphasia. These encephalopathies are characterised in that the spinal cord is not affected as in PVE. The histologic examination reveals vascular lesions and cerebral oedema, with a moderate lymphocytic reaction in the meninges and perivascular spaces and destruction of the lymph node cells. Particularly striking is the demyelination process.

The cerebrospinal fluid is normal with a slight increase in proteins in some instances.

The differences that exist between these forms of neurologic complications depend upon the patient's age: PVE does not affect children under the age of two and, with rare exceptions up to the age of three. In the case of a low immunity, however, encephalopathies may appear with the same incidence in older patients and in children under two. These differences in the manifestations at the level of the central nervous system (CNS) have been attributed to the different response of this system in the various developmental stages. The incidence of encephalopathies under the age of two may be due to the sensitivity of the child to toxic affection of the CNS in any form of severe disease.

Among the neurological manifestations less often encountered are paraesthesias, polyneuritis and meningeal phenomena.

Other complications. Apart from the ocular complications mentioned at autoinoculation, viraemia of the vaccine virus may cause localisation in the eye if there are pre-existing inflammatory eczematous lesions; in general, this occurs in persons with a low resistance, and the aspect is that of a keratitis.

As in other viral diseases (polio, Coxsackie, etc.) viraemia occurring after administration of the vaccine virus may determine cardiac localisations, with myocardial lesions; however, postvaccinal interstitial myocarditis is seldom encountered. In adults, precordial pains, a sensation of constriction, palpitations, dull heart sounds, weak pulse and even rhythm disturbances have been recorded; pericarditis and pulmonary embolism have likewise occurred. In general, these symptoms appear between 8 to 12 days after vaccination. The most plausible of the hypotheses appears to be the direct intervention of the vaccine virus, the interstitial lesion being characteristic of viral infections (Avondet, 1963; Finlay-Jones, 1964).

The vaccine virus may also be located at the level of the osteo-articular system, with suppurative lesions or inflammatory suffusions.

The harmful effects of viral infections in the course of pregnancy, especially during the first three months, are well known, and, in general, smallpox vaccination should be avoided in this situation. The risk of spontaneous abortion is greater in the first trimester, but prematurity, still-birth and the risk of congenital malformations do not appear to be favoured by vaccination, and transmission of vaccinia to the fœtus is an altogether exceptional event.

The prevention of postvaccinal complications consists of:

(i) vaccination at a corresponding age;

(ii) periodical vaccination in order to maintain a sufficient state of immunity;

(iii) careful screening of the subjects with contra-indications, i.e. children with organic defects and a past history of repeated infections. Laboratory tests have shown that these children are immunologically immature and with a plasma cell and gammaglobulinemia deficiency.

It should be stressed that it is unwise to vaccinate children who are hypersensitive or who have eczematous processes, irrespective of whether they are active or not. Moreover, these children should not come in contact with vaccinates in order to avoid heteroinoculation. In the case of revaccination all these children as well as those that have exceeded the age limit must be carefully selected so as to administer the vaccine only to those previously vaccinated.

The contra-indications of smallpox vaccination must be viewed in terms of the epidemiological situation:

(i) in regions of endemic smallpox and especially during epidemics only acute diseases represent a categorical contra-indication;

(ii) in the intact countries where the risk is minimal the sphere of contraindications is much broader.

The contra-indications may be considered as categorical or formal in acute or subacute infectious diseases, in malignant diseases, severe chronic cardiovascular or renal diseases, uncompensated diabetes, neurological and dermatological diseases, severe allergic conditions, dysglobulinemias.

The following are regarded temporary contra-indications: pregnancy, steroid treatment, another recent or concomitant vaccination.

Vaccination may sometimes be put off in the epidemic foci of other viral diseases, especially neurotropic (measles, german measles, poliomyelitis, viral encephalitis), scarlet fever, whooping cough, under weight and premature infants, secondary syphilitic lesions.

Certain chemical compounds (Marboran) and specific human antivaccinal immunoglobulins have been introduced in the treatment of severe complications, the immunoglobulin treatment being assumed to block the circulating virus that produces the complications (Turner, 1962; Adels and Oppe, 1966).

Finally, another means of preventing complications consists in modifying vaccination, either by using an attenuated vaccine or by inducing a state of initial resistance with a killed vaccine (treated with formaldehyde or exposed to ultraviolet light).

4.2.1.7. *Other aspects of smallpox vaccination*

Apart from the technical aspects mentioned, several data concerning the organisation of smallpox vaccination have to be discussed. We refer especially to the potency of the vaccine used, mode of administration, association of the smallpox vaccine with other immunising products, the calendar of smallpox immunisation and the expected results with a carefully organised programme.

Concerning the *potency* of the vaccine, it should be recalled that fewer infective units may be used at vaccination than at revaccination, i.e. 10^7 PFU ml^{-1} being efficient for vaccination and 10^8 PFU ml^{-1} for revaccination. In many countries, India included, the same vaccine is used in both cases.

The WHO recommends *administration* of the vaccine by a single stroke so as to lower the risk of general postvaccinal reactions, and two strokes only for the vaccination of contacts. When vaccination by two strokes is performed the incidence of febrile postvaccinal reactions is somewhat greater. On the other hand, the response in neutralising antibodies is higher and lasts longer (Nyerges *et al.*, 1973). The epidemiological data, however, confirm that a successful vaccination even after a single insertion insures good protection for 4 to 5 years.

In India, the opinion of the specialists differs: they inoculate the vaccine (for primary vaccination and revaccination) in two strokes both by the multiple pressure method and by scarification (with the jet injector a single injection is used). This opinion is based on the fact that in India a single stroke does not insure a corresponding protection, cases of variola occurring among the persons immunised in this way. However, for the primary vaccination in adults and older children a single stroke is used in India, too.

Association of the smallpox vaccine with other immunising products is carried out by simultaneous administration or by combining and administering them together.

The former method was used according to various formulae (smallpox and BCG vaccination, smallpox and TAB vaccination, smallpox and tuberculin testing, etc.).

Simultaneous administration of the smallpox and yellow fever vaccine was already practised before the Second World War in some African countries. The neurotropic yellow fever vaccine was used and applied by scarification.

After introduction of the jet injector the association of vaccines was generally applied in the vaccination campaigns in Africa; moreover further combinations were used:

(i) smallpox vaccine + measles vaccine;

(ii) smallpox vaccine + measles vaccine + yellow fever vaccine (17 D);

(iii) smallpox vaccine + measles vaccine + tetanus vaccine (used as a solvent for the other two);

(iv) smallpox vaccine + measles vaccine + yellow fever vaccine + tetanus anatoxin;

(v) smallpox vaccine + measles vaccine on one arm and BCG vaccine and tetanus anatoxin on the other.

These associated vaccinations were very well tolerated and the immunological response, checked by determining the antibody level, was not significantly modified, irrespective of whether the smallpox vaccine was associated with live or inactivated virus vaccines (measles, yellow fever).

The results obtained show that today a routine or emergency vaccination campaign with multiple objectives and a high efficiency and low cost can be performed.

Nevertheless, most specialists prefer to administer the smallpox live virus vaccine separately in order to avoid postvaccinal accidents (in India it is recommended to perform even the yellow fever vaccine at least 3 weeks after the smallpox one).

The instructions in some countries stipulate that no vaccination should be performed 30 days before and after smallpox vaccination.

There is no general agreement concerning the *calendar* of smallpox vaccinations; some uphold that children should be vaccinated in the second or third trimester of their first year and others in the second year. Arguments have been brought both as regards the period in which the maximum proportion of positive results was obtained and the age at which the proportion of postvaccinal complications was minimal.

As a rule, revaccination is performed at the age of 6 to 7 years when the children begin school, and then at variable intervals: in intact countries when youths are called up for their military service or more often in countries with endemic smallpox (every three years in general and those exposed to infection, the medical staff for instance, every year).

Even in intact countries the personnel of airports and ports, tourist offices, hotels, etc. that come in contact with persons coming from endemic countries, are vaccinated more often.

In Calcutta, children are vaccinated either before leaving the maternity hospital or at the age of 3 to 6 months. When the child is vaccinated within the first few days after birth he is revaccinated at 18 months as the effect might be attenuated by the maternal antibodies; in other cases the children are revaccinated at the age of three.

In India revaccination is repeated in the first and the last year at school, on taking up a job or whenever necessary (in case of family contacts when more than six months have passed since the last vaccination, and in indirect contacts or occupational risks when more than 1 to 3 years have passed).

In general, the Indian specialists consider that protection is certain within the first six months after vaccination, or revaccination, and that it can be relied upon even if there has been direct contact.

If the risk of infection is very great (direct family contact, etc.) or if more than six months have passed after the last vaccination, then immunity is reactivated by revaccination.

Those who are permanently exposed to infection — the medical staff, nurses, auxiliary personnel are vaccinated annually.

The WHO recommends vaccination of the population in endemic countries every three years, as well as of persons travelling in endemic countries. This is considered the maximum interval between vaccinations; in terms of the risk of exposure it may be reduced but not to less than six months.

In intact countries a systematic vaccination programme (first or second year, at the ages of 7 and of 18) is considered sufficient for ensuring immunisation.

For prevention, however, the personnel working at points of international traffic (frontier points, customs, railway lines, airports and ports, hotels, restaurants, tourist offices, buses, etc.) must be vaccinated every three years.

According to the regulations in force in Romania, anybody travelling abroad, except to North America and Europe must have been vaccinated against smallpox within the past three years.

The *value of smallpox vaccination* cannot be questioned; eradication and maintenance of this situation in numerous countries has been achieved by a rigorous vaccination programme. Complete eradication is based upon an increasingly greater proportion of vaccinations in the countries in which smallpox is endemic.

Postvaccinal immunity develops as a rule between the ninth and twelfth day after a successful vaccination and lasts 5 to 7 years. The WHO considers the international vaccination certificate as valid for 3 years, in order to be certain of prevention. Immunity is not however definitive and has to be maintained by periodical vaccinations, the interval between them depending upon the local epidemiological situation.

(1) The efficiency of vaccination is demonstrated by the previously mentioned statistics (cases among vaccinated and non-vaccinated contacts in terms of the number of vaccinations or the interval after the last vaccination) based upon data from endemic countries.

(2) Even when they contract the disease the vaccinates have much milder forms and the mortality rate is much lower. In the New Delhi Hospital the number of deaths among the non-vaccinated children aged one to four was twice that of the vaccinated children.

(3) These data show that vaccinates may eventually contract the disease; in terms of the interval elapsing after the primary vaccination the risk of smallpox is:

(i) 1:1000 after 1 year;
(ii) 1:200 after 3 years;
(iii) 1:8 after 10 years;
(iv) 1:2 after 20 years.

Although the risk of contracting the disease is fairly good the mortality rate is very low even many years after vaccination. When revaccination is regularly performed the resistance even to massive infection is very great.

(4) Emergency revaccination on the outbreak of an epidemic, in a community vaccinated many years before is far more efficient — although the acquired immunity against smallpox has almost

completely disappeared — than in a community lacking immunisation. In India it has been demonstrated that an epidemic can be arrested by revaccination. In contrast, mass vaccination of a totally receptive population in the course of an epidemic has not the expected efficiency and may, according to Buchwald (1965), even give severe forms of the disease in persons receiving the vaccine during the last few days of incubation. This is illustrated by Buchwald from numerous cases.

(5) The efficiency of vaccination during incubation depends upon the moment at which it is performed (the sooner the better the results) and especially upon the degree of infectivity of the patient at the moment of contact.

In general, it is considered that vaccination in the incubation period produces attenuation of the disease and curtails incubation; in Madras the incidence among contacts is as a rule 40%; vaccination performed after contact lowers the incidence to 28%.

(6) These findings determined the authorities in many countries to adopt a rigorous vaccination programme. In other countries, however, even in England in which smallpox vaccination was initiated, it is not compulsory.

This attitude is accounted for by the severe postvaccinal complications that may develop, especially generalised vaccinia and postvaccinal encephalitis. The incidence of the latter differs, as already shown, from one country to another; it appears to be more frequent the older is the child at the time of the primary vaccination. In West Germany the incidence of PVE was 1 per 11 705 subjects during the 1950—1959 period, in the Netherlands 1 per 10 000.

In Austria, according to the statistics of Regamey (1965), the frequency of PVE depends upon the age at which the primary vaccination was performed:

(i) from 1—3 years 1 : 10,000;
(ii) from 4—5 years 1 : 650;
(iii) from 11—14 years 1 : 120.

If, formerly, the high number of postvaccinal complications and deaths were justified in view of the high mortality rate of variola, today this can no longer be admitted in the intact countries.

The attempts at preparing killed virus vaccines: the formaldehyde treated vaccine (Herrlich, 1958, 1964) or inactivated by ultraviolet light did not give the expected results; the use of highly purified soluble antigens was also recommended. Even if they produce antibodies, the killed virus vaccines do not confer a lasting corresponding immunity and may sensitise the vaccinate. Consequently, the use of an attenuated, slightly antigenic strain (Rivers) was resorted to in the form of the PDN vaccine (Posthuma, Dekking, van der Noordaa). These vaccines have been recommended in cases of emergency, in non-vaccinated adults or other contra-indications, in order to prepare the individual for inoculation with the standard vaccine.

In these situations, however, vaccination under the protection of specific antivaccinal immunoglobulin or N-metlylisatine-thiosemicarbazone is preferred to prevaccination (Kempe, 1961 ; Marennikova, 1961 ; Jaroszynska-Weinberger and Meszaros, 1966 ; Adels and Oppe, 1966).

(7) At present the problem of the opportunity of maintaining a compulsory or even routine smallpox vaccination programme in the intact countries has stimulated controversy in view of the evolution of variola throughout the world. Even if the number of cases has increased annually, the number of countries with smallpox cases gets smaller and smaller. With progress in the eradication programme in the endemic countries, the potential danger of importing the disease in the intact countries is much reduced, particularly when the frontier health regulations are complied with.

The smallpox case imported in Japan in March 1973 remained isolated without causing any further case.

Many states have reviewed their attitude towards routine vaccination. Thus, in England, smallpox vaccination is not routinely applied ; it is only compulsory for those who leave for, or come from, endemic countries (or in which the eradication programme is being applied). Vaccination is, however, compulsory for the medical personnel coming in contact with the patients. A similar attitude has been adopted in the U.S.A.

The WHO considers that in the countries with a well organised health network the fact that systematic vaccination has been given up only represents a restricted risk, since any epidemic episode is rapidly detected and eradicated ; where the sanitary organisation is not highly efficient, however, the consequences of renouncing vaccination may spell disaster.

The elimination of routine vaccination demands therefore a perfect organisation of the health service which implies rapid action for the detection of any imported case, and of the primary, secondary or indirect contacts. In addition, the elimination of routine vaccination demands a corresponding supply of vaccines, chemical substances and equipment so as to be able to intervene operatively at any moment.

In conclusion :

(i) vaccination is the only method for the prevention and control of smallpox ;

(ii) routine vaccination and periodical revaccination insures a high level of resistance among the population;

(iii) in a declared focus vaccination lowers the receptivity of the population to the disease, especially when it has already been in contact with the vaccine virus;

(iv) for vaccination, performed according to the most efficient techniques, an efficient vaccine must be used, transported, stored and handled in the best conditions;

(v) correct recording of the results and repeated vaccination in the negative or equivocal cases raises the proportion of protected subjects;

(vi) screening of the non-vaccinated and contra-indicated cases avoids severe complications;

(vii) when the danger of infection increases, vaccination must be extended to groups of the population which are usually exempt from vaccination, and must be carried out under the protection of specific immunoglobulin or drugs;

(viii) periodical revaccination at adequate intervals of the persons exposed to the virus significantly lowers the risk of infection.

4.2.1.8. *Aspects of the organisation of vaccination*

The *smallpox vaccine* may be prepared from calves and delivered in both fluid and lyophilised form.

The *fluid vaccine* is a glycerolated virus vaccine suspension in vials (for approximately 50 vaccinations) and in individual capillary tubes; the content of the vial is turbid and must be shaken before use.

The vaccine is stored in the dark at temperatures of +4° to +15°C; the duration of its efficiency is marked upon the label.

The *lyophilised vaccine* is, as a rule, supplied in 5 ml vials, containing 0.5 ml vaccine virus suspension in buffer phosphate with peptone (5%), phenol (0.4%) and neomycin.

Kept at +4°C, the lyophilised vaccine is efficient for two years. To the content of the 5 ml vial, 0.6 ml glycerine (50%) in distilled water is added; the vaccine can then be used within an interval of 30 days provided it is kept in the dark at +4°C.

Fluid vaccine may be used for both the primary vaccination and revaccination, the lyophilised one preferably for revaccination.

Vaccination is recommended between the ages of one to two years and is followed by revaccination in general at the age of six to seven years. A second revaccination is carried out when the youths reach maturity. Finally, revaccination is also performed whenever the possibility of contamination exists (international travelling, or the personnel assigned to international traffic) as well as in special epidemiologic situations.

It is recommended that smallpox vaccinations should be performed in April-May and September-October.

Vaccination by scarification is performed on the skin of the left arm at the level of the insertion of the deltoid and revaccination on the right arm. It is recommended:

(i) to wash the vaccination area with soap and water;

(ii) to dry the skin;

(iii) to sterilise the needle or stylet by heat and cool it before applying the vaccine.

Scarification must not go deeper than 6 mm and should be performed through the vaccine put on the skin. Sanguinolent traces must appear within 30 seconds; the vaccine is then pushed in with the edge of the instrument. The site of inoculation is then left to dry without exposing the arm to the sun or heat. Clean linen should then be put on.

Vaccination is performed by a doctor or under his direct surveillance.

The results are read 6 to 8 days after vaccination.

The data must be recorded in detail (successful vaccination or failure, local or general reactions, complications).

A successful primary vaccination is marked by the typical reaction already described:

(i) major reaction = successful vaccination (+) (vesicular or pustular lesion or a congestive or indurated zone surrounding an ulcer or central scab);

(ii) equivocal reaction = any other response (+);

(iii) negative reaction = no kind of lesion (−).

Immunisation is repeated in case of an equivocal or negative reaction (in case of revaccination) and the result read after 8 days.

Persons not immunised up to the age of two years will only be vaccinated in special epidemiological situations and only under the protection of specific immunoglobulin (0.3 ml kg^{-1} body weight, but not more than 10 ml).

As a rule only persons with a characteristic cicatrix, proof of a previous vaccination, will be inoculated.

For the primary vaccination the contra-indications established are temporary (febrile diseases, acute digestive disorders, severe nutritional diseases, persons in contact with cases of communicable diseases in the incubation period, acute pruriginous affections of the skin, cortisone or immunosuppressive treatment; healthy children who live with persons suffering from eczema and who cannot be isolated for 20 days after vaccination are likewise exempted) or permanent (chronic pruriginous conditions, eczemas, neurological diseases, children with nervous defects, malignant haemopathy, haemorrhagic syndromes, cancer, dysgammaglobulinaemia).

On revaccination, pregnancy (irrespective of the age), respiratory, cardiac, renal or hepatic insufficiency must be added to the list of contra-indications.

When vaccination is absolutely necessary because of an exceptional epidemiological situation then it should be performed under the protection of specific immunoglobulin.

4. 2. 2. Other preventive measures

Bearing in mind the possibility of an imported smallpox case the medico-sanitary personnel must be instructed in due time concerning the clinical picture of smallpox (especially the atypical forms), the laboratory diagnosis and anti-epidemic measures to be taken in the focus. This includes the personnel at the units where the first cases of smallpox may be encountered; that is, the frontier points (airports and sea ports), ambulance services, polyclinics in the towns connected with international traffic or tourism, communicable diseases hospitals. This personnel must be fully acquainted with the measures to be taken regarding a variola case and the contacts.

The hospital unit that will be assigned to the care of eventual smallpox cases or quarantine of the contacts has to be established in advance, as well as the means of evacuation. This implies operativity and no risk of infection for the staff members or patients in the hospital within which the smallpox unit will be.

A plan for the prevention and control of variola must be drawn up, also establishing the personnel that will be called up, their activity and the necessary equipment and materials.

The personnel that will be called upon to work from the very first cases must be vaccinated every three years and fully instructed in anti-epidemic measures, care of the patient, etc.

Among the preventive measures of particular importance is the compulsory revaccination of foreign travellers coming from endemic areas or focuses (if not vaccinated during the last three years) and application of the International Health Regulations.

4. 2. 3. Control measures

The prevention of an epidemic when one or several cases appear in an intact country depends upon early detection of the patient and contacts, their strict and rapid isolation, and active immunisation, within as short an interval as possible, of the entire population within the exposed area.

4.2.3.1. *Measures concerning the patient*

Early detection of patients necessitates active survey of the area and instruction of the personnel in order to recognise smallpox in all its forms of manifestation.

Any suspect case is notified by telephone, normally, within the first hours of detection and *isolated* in a smallpox unit.

Isolation implies strict compliance to the rules in order to prevent transmission of the disease within the hospital or outside it.

The patient will be kept in hospital until complete clinical recovery (fall of the last scabs). Before discharge he will be given prolonged baths with antiseptics, paying particular attention to disinfection of the scalp.

As the variola virus is extremely resistant to the medium, *disinfection* plays a major role in the control of smallpox. Disinfection must have in view complete destruction of the virus in the patient's clothes, linen and bedclothes, all objects coming in contact with him the rooms and the pathologic products (mucous secretions, scabs).

Disinfection must include:

(i) the house and working site of the patient and contacts;
(ii) any place where the patient has been working;
(iii) the car in which he was transported;
(iv) the ward (hospital or unit) where he was isolated.

Disinfection must be done within the first hours after detection and isolation of the patient; current disinfection must by performed in hospital and a final disinfection after discharge of the patient.

There are different *methods of disinfection*. The most efficient is burning, a method that can be applied to newspapers, books, mattresses, certain clothes, rags, etc.

When disinfectants are not available the bedclothes, linen and clothes can be boiled; the walls and floor can be washed with soap and water and left to dry for the next 48 hours, letting the sun in as much as possible.

The following disinfectants are recommended:

(i) formaldehyde in a concentration of at least 1%;

(ii) phenol compounds;

(iii) quaternary ammonium compounds in a concentration of at least 1% in which various fabrics and objects are soaked for 2 to 24 hours.

The rooms are disinfected by spraying the disinfectant or by washing the floor, walls, doors, etc. with it; a contact of at least 4 hours must be insured. Disinfection is then followed by washing with water. In case of disinfection with formaldehyde vapour (an excellent method for the disinfection of rooms) the contact must be of at least 6 hours.

It is recommended to boil the commercial formaldehyde in two volumes water (500 ml formalin plus 1 liter water, for 30 cm^3) or to add potassium permanganate to formalin (150—200 g potassium permanganate for 500 ml formalin in 1 litre water for 30 cm^3).

The linen, clothes etc. are kept in a disinfectant solution for 2 to 24 hours.

A strong penetrating disinfectant is ethylene oxide; its use involves special apparatus and skilled personnel.

When disinfection is performed in hospital, at home, at the site of work, no object with which the patient has come in contact must be omitted or removed before disinfection.

In no case must the carpets, clothes, bedlinen, etc. be shaken out of the window or out of doors.

The scabs that fall must be collected and burnt and the container disinfected.

4.2.3.2. *Measures concerning the contacts*

The contacts are kept in quarantine for 21 days from the moment in which they are considered to have come in contact. Isolation in hospital may be given up in view of the danger of in-hospital transmission of the disease either from a confirmed patient or from a contact who has contracted the disease. The contacts may be kept in quarantine at home or in suitable rooms (in motels, chalets, etc.) far from populated centres (as in Yugoslavia during the 1972 epidemic).

During quarantine the contacts must be under daily medical survey. Their temperature is recorded every day.

The contacts are immediately vaccinated since it is known that the longer the interval between infection and application of the vaccine the fewer are the chances of avoiding the disease. Maximum chances exist when vaccination is done within 24 hours after coming in contact with a smallpox case.

Rapid identification of all the contacts for their isolation and vaccination depends upon the quality of the epidemiological survey.

During the last few years promising results have been communicated concerning protection of family contacts by means of antiviral drugs. Among the most efficient chemoprophylactic drugs tested is *N*-methylisatine-ß-thiosemicarbazone (Ribeiro do Valle *et al.*, 1964); it has the disadvantage, however, of producing nausea and vomiting in a comparatively high proportion of the persons to whom it is administered (Bauer *et al.*, 1969).

The WHO and health officers in India underline the fact that vaccination as a preventive measure cannot be replaced by chemoprophylactic drugs, which may eventually be used for the protection of family contacts that have not been previously vaccinated or those with major contra-indications. Similarly, antivaccinal immunoglobulins, which also prevent postvaccinal complications, may be administered to the non-vaccinated family contacts.

4.2.3.3. *Measures of prevention regarding the receptive population*

Immediately a smallpox focus is declared the entire population in the exposed area is vaccinated (primary or re-vaccination) using an efficient vaccine so that the takes should be of at least 95%. Vaccination should be repeated in those with an equivocal or negative reaction within the first five days but only on approval of health officials.

References

Adels B. R., Oppe F. E. (1966) *Lancet*, *1*, 7427.

Arita I., Henderson D. A. (1968) *WHO Bull.*, *39*, 277.

Avakian A. A., Alstein A. D., Kirilova F. M., Bikovski A. F. (1961) *Vopr. Virusol.* (Moscow), *6*, 2.

Avondet A. (1963) *Méd. Hyg.* (Genève), *21*, 765.

Babeş V. T., Kimishkian Cl. (1972) *Rev. Roum. Virol.*, *9*, 85.

Balş M., Căruntu Fl. (1965) *Rev. Roum. d'Inframicrobiol.*, *2*, 187.

Baroian O. N., Serenko A. F. (1961) *J. Microbiol. Epidemiol. Imunol.* (Moscow), *4*, 72.

Bauer J., Sadler P. (1960) *Lancet*, *1*, 110.

Bauer J., St. Vincent L., Kempe C. H., Young P. A., Downie A. W. (1969) *Amer. J. Epidemiology*, *90*, 130.

Baylet R., Collomb H., Grelier L. (1963) *Rev. Neurol.* (Paris), 108.

Benda R., Noskov F. S., Danes L., Serbezov V., Goldin R. (1968) *Acta Virologica*, *12*, 229.

Buchwald G. (1965) *Die Med. Welt*, *4*, 65.

Cajal N. (1958) *Diagnosticul de laborator al inframicrobiozelor umane (The Laboratory Diagnosis of Human Viral Diseases)* Ed. Academiei R.S.R., Bucureşti.

Cruickshank J. G., Bedson H. S., Watson D. H. (1966) *Lancet*, *2*, 7462.

Dixon C. W. (1962) *Smallpox*, J. and A. Churchill Ltd., London.

Downie A. W., Meiklejohn G., St. Vincent L., Rao A. R., Sundara-Babu B. V., Kempe C. H. (1965) *WHO Bull.*, *33*, 623.

Dumbell K. R., Nizamuddin M. D. (1959) *Lancet*, *2*, 916.

Dumbell K. R., Bedson H. S., Rosier E. (1961) *WHO Bull.*, *25*, 73.

Dumbell K. R. (1968) *Laboratory Aids to the Control of Smallpox in Countries where the Disease is not Endemic*, in *Progress in Medical Virology*, 10, edited by J. L. Melnick, S. Karger, Basel-New York.

El-Ganzoury A.L.A. (1967) *J. Clin. Path.*, *20*, 879.

Eremian A. V. (1962) *Vopr. Virusol.* (Moscow), *5*, 597.

Fenner F. (1949), *J. Immunol.*, *63*, 341.

Fenner F. (1968) *The Biology of Animal Viruses*, Academic Press, New York-London.

Finaly-Jones L. R. (1964) *New Engl. J. Med.*, *270*, 41.
Gordon J. D., Fothergill K.F.L., Millar E.L.M., Flewett T. H., Bedson H. S., Cruickshank J. G. (1966) *Lancet*, *1*, 1311.
Gray I. (1949) *Lancet*, *1*, 749.
Grist N. R. (1968) *Health Bull.*, *XXVI*, 20.
Helbert C. (1957) *Lancet*, *1*, 1012.
Henderson A., Arita I. (1973) *WHO Chronicle*, *27*, 155.
Herrlich A. (1952) *Deutsche Med. Wschr.*, *87*, 71.
Herrlich A. (1958) *Die Med. Welt*, *34*, 1426.
Herrlich A. (1964) *Deutsch. Med. Wschr.*, *189*, 968.
Jaroszynska-Weinberger B., Meszaros J. (1966) *Lancet*, *1*, 948.
Johnes B. R., Huseine M. K. (1963) *Trans. Ophthalm. Soc.* U.K., *83*, 613.
Joklik W. K. (1968) in *Molecular Basis of Virology* by Fraenkel-Conrat H., Reinhold Book Corporation, New York-Amsterdam-London.
Kempe C. H. (1961) *WHO Bull.*, *25*, 71.
Kempe C. H., St. Vincent L. (1964) *Variola and Vaccinia Viruses in Diagnostic Procedures for Viral and Rickettsial Diseases* by E. H. Lennette and N. S. Schmidt, Amer. Public Health Assoc., 3rd ed.
Lépine P. (1964) *Techniques de laboratoire en virologie humaine*, Masson et Co., Paris.
Mac Callum F. O., Mc Donald J. R. (1957) *WHO Bull.*, *16*, 247.
Marennikova S. S. (1961) *J. microbiol., epidemiol., imunol.* (Moscow), *4*, 84.
Marennikova S. S., Malteva N. N. (1961) *Vopr. Virusol.*, *6*, 204.
Metha B. C., Purandare N. M., Patel J. C. (1967) *Indian Journal of Medical Sciences*, *21*, 518.
Millar J. D., Roberto R. R., Wulff H., Venner H. A., Henderson D. A. (1969) *WHO Bull.*, *41*, 749.
Moustardier G. (1973) *Virologie Médicale*, Ed. Maloine, Paris.
Murray H.G.S. (1963) *Lancet*, *1*, 847.
Nagler F. P. O., Rake G. (1948) *J. Bact.*, *55*, 45.
Neff J. M., Millar J. D., Roberto R. R., Wulff H. C. (1969) *WHO Bull.*, *41*, 771.
Netter R., Lapeyre D. (1964) *Rev. Hyg. Med. Soc.*, *12*, 475.
Netter R., Sibilat A. M. (1968) *WHO Bull.*, *39*, 940.
Nicolau St. S., Cajal N. (1962) *Elemente de virologie (Elements of Virology)*, Ed. Academiei R.P.R., 395.
Nicoli J., Jolibois C. (1964 a) *Ann. Inst. Pasteur*, *107*, 374.
Nicoli J., Jolibois C. (1964 b) *Ann Inst. Pasteur*, *107*, 453.
Nizamuddin M. D., Dumbell K. R. (1961) *Lancet*, *1*, 68.
Noble J., Rich J. A. (1969) *WHO Bull.*, *40*, 279.
Nyerges G., Erdos L., Melly F. (1973) *WHO Bull.*, *48*, 397.
O'Connell C. J., Karzon D. T., Barron Al. L. (1964) *Ann. Int. Med.*, *60*, 282.
O'Sullivan D. G., Sadler P. W., Russel V. (1964) *Arch. Ges. Virusforsch.*, *14*, 650.
Pickford Marsden J. (1962) *Lancet*, *5503*, 524.
Rake G., Blank H., Carriell L. L., Nagler F.P.O., Scott T.F.M. (1948) *J. Bact.*, *56*, 293.
Rao A. R. (1964) *Ind. Med. Assoc.*, *43*, 224.
Rao A. R. (1966) *Bull. Ind. Soc. Med. Com. Dis.*, *3*, 2.
Rao A. R., Mc Fadzean I. A. (1966) *Lancet*, *1*, 1068.
Ratner S. I., Hudeakova G. K. (1961) *Klin. Med.*, *4*, 51.
Regamey R. H. (1965) *Rev. Intern. des Services de Santé*, *38*, 269.
Ribeiro do Valle L. A., de Melo P. R., de Salles Gomes L. F. (1965) *Lancet*, *2*, 976.

Ricketts T. F., Byles I. R. (1908) *Diagnosis of Smallpox*, Cassell, London.
Roberto R. R., Wulff H., Millar J. D. (1969) *WHO Bull.*, *41*, 761.
Rodriguez C. L., Santos R. S. (1965) *Rev. Intern. des Services de Santé*, *38*, 301.
Sarkar J. K. (1964) *Indian Med. Assoc.*, *43*, 343.
Sarkar J. K., Mitra A. C., Mukerjee M. K., De S. P. (1973 a) *WHO Bull.*, *48*, 523.
Sarkar J. K., Mitra A. C., Mukerjee M. K., De S.P., Mazundar D. G. (1973 b) *WHO Bull.*, *48*, 517.
Turner W. (1962) *Brit. Med. J.*, *1*, 1317.
Vichniakov V. E. (1968) *WHO Bull.*, *39*, 433.
Vieuchange J., de Brion G., Gruest J. (1958) *Ann. Inst. Pasteur*, *95*, 681.
Wolf H. L., Croon J.J.A.B. (1968) *WHO Bull.*, *38*, 492.
Zwartouw H. T., Wertwood J.C.N., Harris W. J. (1965) *J. Gen. Microb.*, *38*, 39.

The authors of the present work have read some other studies, which they recommend to the readers, as follows :

Briody B. A., *Pox Virus* in *Basic Medical Virology*, edited by Prier J., The Williams & Wilkins Company, Baltimore, 1966.
Britton G.S.C., *Disorders of Blood*, Ed. J. Churchill Ltd., London, 1956.
Brown E. N., *Complications of Smallpox Vaccination*, Med. J., 1965, *41*, 734.
Cajal N., *Diagnosticul de laborator al inframicrobiozelor umane (The Laboratory Diagnosis of Human Viral Diseases)*, Ed. Academiei R.P.R., București, 1958.
Comité d'experts de la variole, *Premier Rapport*, OMS, Série de rapports technique, no. 493, Genève, 1972.
Comité d'experts de la variole, *Deuxième Rapport*, OMS, Série de rapports technique no. 493, Genève, 1972.
Eradication de la variole, OMS, Série de rapports technique, no. 393, Genève, 1968.
Dixon C. W., *Smallpox*, Ed. J. and A. Churchill Ltd., London, 1962.
Dumbell K. R., *Laboratory Aids to the Control of Smallpox in Countries where the Disease is not Endemic*, in *Progress in Medical Virology*, 10, edited by J. L. Melnick, S. Karger, Basel-New York, 1968.
Fenner F., *The Biology of Animal Viruses*, Academic Press, New York-London, 1968.
Gracias R., *An Introduction to Diagnosis, Treatment and Control of Infectious Diseases*, St. Patrick's Press, Bishop's House, Poona, 1967.
Joklik W. K., In *Molecular Basis of Virology* by Fraenkel-Conrat H., Reinhold Book Corporation, New York-Amsterdam-London, 1968.
Kempe C. H., St. Vincent L., *Variola and Vaccinia Viruses in Diagnostic Procedures for Viral and Rickettsial Diseases* by E. H. Lennetie and N. S. Schmidt, Amer. Public Health Assoc. Inc., third edition, 1964.
Kempe C. H., Benenson A. S., *Smallpox Vaccination in the United State*, J.A.M.A., 1965, *194*, 2, 161.
Le diagnostic de la variole au laboratoire, OMS, Genève, 1969.
Les maladies transmissibles en 1969, Chronique OMS, 1970, *24*, 6, 284.
Lépine P., *Techniques de laboratoire en virologie humaine*, Masson et Co., Paris, 1964.
Mac Callum F. O., *The Dynamics of Virus and Rickettsial Infections*, chap. 24, Ed. Balkiston, New York, 1964.
Moustardier G., *Virologie Médicale*, Ed. Maloine, Paris, 1973.
Nestorescu N., Strati I., Popescu Al., *Achiziții recente în diagnosticul de laborator*

al unor boli deosebit de severe (Recent Progress in the Laboratory Diagnosis of Some Particularly Severe Diseases), Revista Sanitară Militară, 1966, *LXII*, 5, 809.
Nicolau St. S., Cajal N., *Elemente de virologie (Elements of Virology)*, Ed. Academiei R. P. R. Bucureşti, 1962.
Où en est l'éradication de la variole, Chronique OMS, 1970, *24*, 7, 329.
Propagation de la variole par l'air, Chronique OMS, 1970, *24*, 7, 339.
Rél. épid. hebdomadaire, OMS, 1969, *44*.
Rél. épid. hebdomadaire, OMS, 1970, *45*.
Rél. épid. hebdomadaire, OMS, 1971, *46*.
Rél. épid. hebdomadaire, OMS, 1972, *47*.
Rél. épid. hebdomadaire, OMS, 1973, *48*.
Rél. épid. hebdomadaire, OMS, 1974, *49*.
Ricketts T. F., Byles J. R., *Diagnosis of Smallpox*, Cassell, London, 1908.
Smadel J. E., *Smallpox and Vaccinia* in *Viral and Rickettsial Infection of Man*, edited by Th. Rivers, J. B. Lippincott Company, Philadelphia-Montreal, second edition, 1952.
Strati I., Ştefănescu V., *Stadiul actual al diagnosticului de laborator al variolei (Present Stage in the Laboratory Diagnosis of Smallpox)*, Studii şi cercetări de inframicrobiologie, 1970, *21*, 1, 63.
Tudor V., Strati I., *Probleme actuale legate de vaccinarea antivariolică în forţele armate (Present Problems Linked to Smallpox Vaccination in the Armed Forces)*, Revista Sanitară Militară, 1966, *LXII*, 4, 629.
Tudor V., Strati I., *Unele aspecte actuale ale variolei în India (The Present Aspects of Smallpox in India)*, Viaţa Medicală, 1969, *XVI*, 4, 259.

PART II

Cholera

5. General

Cholera is an acute infectious disease caused by Vibrio coma and characterised by a brutal onset, specific diarrhœa and severe evolution with massive dehydration, collapse and shock.

Its high propagation potential and case fatality rate made cholera the most terrifying disease of the 19th century.

In order to distinguish it from other diseases producing dehydration and collapse it was called *Asiatic cholera*, the south of Asia being its cradle and the main endemic focus.

In the past, Asiatic cholera was confused with the severe forms of alimentary toxic infection *(cholera nostras)* and severe infantile enterocolitis *(cholera infantum)*.

Initially, the incriminated germ was *Vibrio cholerae* with its classical variants *Inaba*, *Ogawa* and *Hikojima*.

Discovered casually in 1905, in the cadavers of pilgrims at the quarantine point of El Tor, *Vibrio El Tor* was ignored and not considered pathogenic until 1961—1962, when it precipitated an epidemic with all the clinical attributes of classical cholera in the Indonesian Islands. During an epidemic in Calcutta, Mukerjee *et al.* (1965) found that 90% of the cases were caused by the El Tor lysotype, which then advanced to threaten other regions from which the disease had disappeared many years ago, even reaching the Soviet Union. By 1962—1966 El Tor had invaded 23 countries.

After a period of quiescence (1967—1969), in 1970 cholera invaded new territories in Asia, passed into Africa and then Europe.

This led the WHO experts to reconsider the El Tor vibrio and list it together with the classical vibrio.

The appearance of the El Tor vibrio is not an aetio-epidemiological novelty, but it nevertheless brought about clinical, therapeutical and prophylactic changes in the cholera focus.

The clinical picture of classical cholera changed, the intermediate and milder forms of the disease becoming predominant with *Vibrio El Tor* present in the stools, to the surprise of various investigators.

It was even assumed that although cholera had reappeared it was milder and no longer fatal as in previous epidemics.

The surprises continued when it was discovered that El Tor gave typical and even severe forms of cholera, with a rapid lethal evolution, that allowed for no therapeutical intervention.

It is very dangerous to underrate the pathogenic attributes of *Vibrio El Tor*. Its rapid spread during the last few years and its appearance in areas in which it was considered a thing of the past, demands far greater attention. Lending support to this point of view is the invasion of Western African countries by cholera.

This situation appears to be favoured by a marked degree of adaptation and resistance to environmental conditions of *Vibrio El Tor* as compared to the classical strains, and by the existence of a great number of atypical and subclinical forms (short or long term carriers) whose outcome leads to vast epidemiological implications.

This is not an isolated fact and may be compared to the similar occurrence with *Shigella shigae* the most feared of the dysentery bacilli, which formerly caused all the severe cases of dysentery throughout the world, and was gradually replaced (a process that is still taking place today) by the so-called pseudodysentery bacilli, belonging to the Flexner, Sonne, Schmitz groups, etc., to which no attention was paid some 20 to 30 years ago.

Change in the aetiological structure of the disease was followed by a change in its clinical aspect, which today includes a wide range of manifestations, starting from the severe forms of dysentery up to the mild atypical and subclinical forms, extending dysentery like an "oil spot" over the world and transforming it into an endemic disease.

As in dysentery caused by *S. shigae*, rarely encountered today, the classical cholera epidemics have become less and less frequent. According to Mukerjee *et al.* (1965), 90% of the cholera epidemics in South Asia are caused by El Tor vibrio.

Irrespective of the vibrio that causes the disease (the near future may bring forth other types) cholera remains a severe disease, with an unforeseeable evolution that exhausts the organism and exposes it even after recovery to severe complications.

5. 1. The history and geographical spread of cholera

In contrast to other infectious diseases that know no geographicas limits, cholera has been known from ancient times to be more or lesl established on the Asiatic continent especially in India and Pakistan and the large deltas of the Ganges and Brahmaputra.

In the age-old evolution of cholera there were moments in which the endemic region spread beyond these areas eastwards and especially westwards, but remained along general lines within the traditional Indopakistan focus until 1817, after which it migrated along the trade routes.

After 1817 there were six pandemics during which few areas escaped: the north of Siberia, part of the Scandinavian countries, Faroe, Shetland and Orkney islands, Hebrides, Lapland, Greenland, Falkland Island, Iceland, Western coast of Africa, Argentine, Chile, Bermudas and probably Australia and New Zealand (Fig. 5.1).

Except for the first, Europe suffered from all the other epidemic waves.

This wide spread of the disease appears to have been due to a great extent to ignorance of the causes and, linked to this fact, to the absence of efficient measures of prevention and control.

It was not until 1883 that the first effective measures were taken, after Koch discovered the bacillus. However, the disease continued to spread owing to the steady development of routes of communication, especially by sea.

During World War I only a few isolated cases were recorded on and behind the front. After the war, starting in 1923 the cholera epidemic gradually faded, returning to its traditional endemic focus which, down the ages, always represented the starting point and only source of the great pandemics. Each time *Vibro cholerae* was isolated.

Between 1923 and 1960 the importance of cholera decreased and except for the epidemics of Iran (1939) and Egypt and Syria (1947—1948) no other countries were invaded.

These new circumstances led to the assumption that the disease would disappear spontaneously and might therefore be eradicated in the near future.

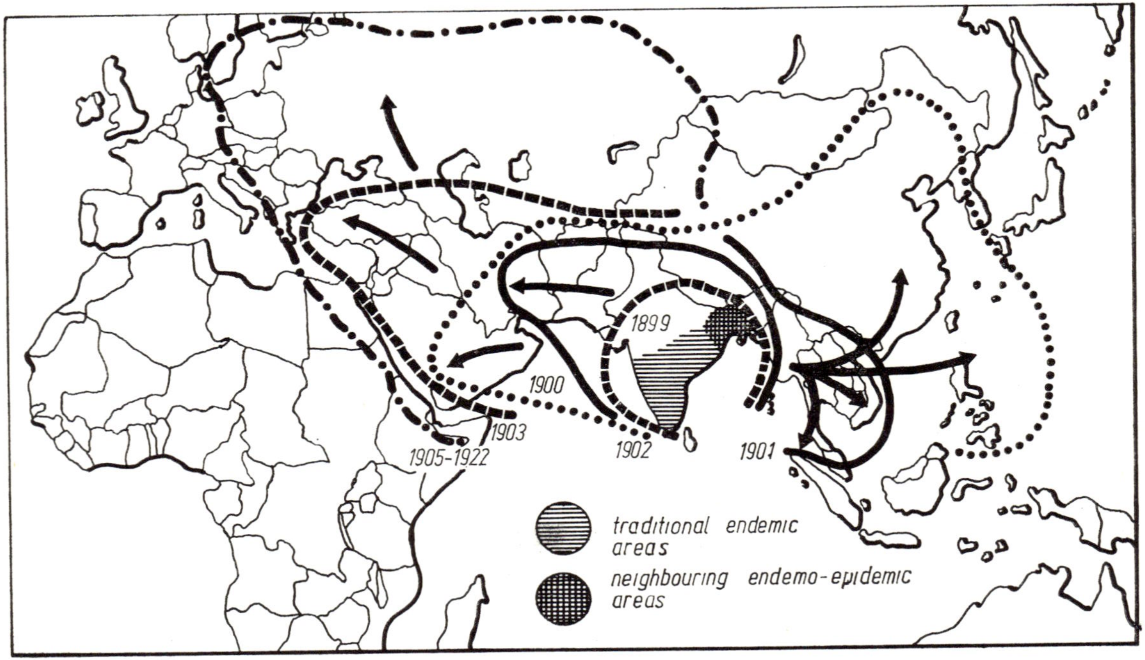

The spread of cholera during the 6th pandemic (1899–1922) (WHO Chronicle, 1970, 20, 7)
FIG. 5.1.

The history of cholera between 1960—1961 marks a new stage, comparable to that of 1817, the disease again taking up its westward trend.

The starting point this time was Indonesia and, surprisingly, the El Tor biotype was isolated and not the classical cholera vibrio.

The history of El Tor begins at the turn of the century when it was first discovered in the *El Tor* lazaretto (on the western coastline of Mount Sinai) serving as a quarantine base for pilgrims returning from Mecca. In 1897, Ruffert isolated a vibrio similar to that of cholera on performing the autopsy on a pilgrim who died suddenly without presenting the symptoms of the disease.

In 1905, Gottschlich, examining the stools of 107 healthy pilgrims returning from Mecca, isolated vibrios agglutinated by cholera antiserum in six cases and considered them carriers. As the pilgrims had left their homes two months previously Gottschlich believed that they had been infected by the holy Zam-Zam water next to the great mosque of Mecca. Thirty years later Abdelrachman isolated from the sources of Zam-Zam a vibrio agglutinated by cholera haemolytic antiserum 0, similar to that isolated by Gottschlich in the El Tor lazaretto. Until then *Vibrio El Tor* had not been accepted as the aetiologic agent of cholera and was considered apathogenic because it had been isolated from individuals who had given no signs of the disease.

In 1937—1938, in the Celebes (the Makassar district) there was an outbreak of a disease with all the characteristic symptoms of cholera. There were several extremely severe forms of cholera — cholera sicca — which causes death within a few hours. The epidemic, 48 cases, lasted from September 1937 until April 1938.

A new epidemic outbreak occurred in October 1939 on the eastern coast of the Celebes and lasted until 1940, totalling 62 cases.

The morbidity was four per thousand and the mortality rate 70%. In both epidemics the El Tor biotype was isolated both from the patients and from the waters of the rivers. In 1944, during the Japanese occupation, there were seven more cases.

These epidemics opened up again the most discussed problem of the role and importance *of Vibrio El Tor* in the aetiology of cholera.

It was assumed that the El Tor biotype was an opportunist bacillus and the disease was labelled *El Tor choleriform enteritis* or *El Tor paracholera.*

Finally, it was postulated that the El Tor biotype had two variants: an apathogenic Hedjaz variant deriving from the Zam-

Zam strain, and a pathogenic Makassar variant responsible for the epidemics of the Celebes. In view of the limited character of the Celebes epidemics and the fact that although agglutinated by the cholera antiserum 0 the isolated vibrios were haemolytic, it was not considered necessary to extend the quarantine measures, applied in cases of classical cholera, to the El Tor paracholera.

For 13 years there was a period of quiescence and it was thought that the appearance of the El Tor vibrio in Celebes had been an epidemic incident without any significance.

At the beginning of January 1957, cholera appeared in the town of Makassar and the neighbourhood and lasted until February the next year. There were 151 cases and two in Djakarta. The case fatality rate was still higher this time – 90%, and the disease remained endemic in the country.

Cholera then passed on to the other islands of Indonesia (Java, Borneo, Sumatra) invading numerous localities in the course of 1969,

On February 1 1962, the epidemic totalled more than 4000 typical cholera cases with almost 900 deaths (21%), all in Indonesia.

The Asiatic endemic continued its course showing a tendency to decrease in Bengal in 1960 but new epidemics broke out in western Pakistan (1958–1960) and Afghanistan (1960), as well as in Thailand (1958–1959).

In Thailand the disease reappeared after an interval of 8 years. From the stool of the first patient in Thonburi, near Bangkok, the cholera vibrio Ogawa serotype was isolated, similar to the vibrio responsible for the concomitant outbreak in Calcutta. The epidemiological survey showed that the epidemic had spread by water from an Indian peasant living upstream of Thailand. The disease was not transmitted by land as there were no cases in Burma, which separates Thailand from Bangladesh.

There were 19 359 cholera cases in Thailand.

An epidemic focus broke out again in western Pakistan, assumed to have been started by a tourist from Bombay. In 1960 there were 6133 cases. The neighbouring Indian states were likewise affected – Punjab, Jammu and Kashmir. In September 1960 the disease passed into Afghanistan.

The epidemiological situation of 1958–1960 did not forecast the extent of the increase in cholera that would take place in the following years. In these cases the classical *Vibrio cholerae* was isolated but it cannot be ascertained that all the cases were caused by this type. At any rate it seems that the endemic focus broke out about the time that the El Tor biotype re-entered Asia.

In September-December 1960, in Ubol (Thailand), there was an outbreak of mild cases, without any deaths, with recovery within 24 to 48 hours after starting treatment. From the stools of the patients the *Ogawa serotype of Vibrio El Tor* was isolated, intensely haemolytic and resistant to phagotype IV. This biotype was isolated both from water and from contaminated food. It was difficult to establish the origin of the epidemic since the epidemic of 1958–1960 had been caused by *Vibrio cholerae.*

In June-July 1961, El Tor cholera passed to Hong Kong, in August into Kwantung, and in September into the Philippines where no case had been recorded for 25 years. In the Philippines the first cases appeared in Manila, among the population of the harbour, then rapidly spread throughout the country, totalling 15 000 cases. In 1962, El Tor cholera passed into Taiwan and South Korea. It became evident that El Tor cholera had taken on all the attributes and potential of a true pandemic (Figs 5.2 and 5.3).

At the XV World Health Session the "El Tor Problem" was taken up and the *Vibrio El Tor* placed next to the classical vibrio, El Tor cholera being embodied in the comprehensive, classical, clinical picture of cholera as an internationally notifiable disease. Application of the international health rules was recommended.

After 1962, El Tor cholera invaded, in turn, all the south-eastern Asiatic states: in 1963 it reappeared in Thailand, Malaya, Cambodia, the Mekong delta, Borneo and Bangladesh. In April 1964 the first case of El Tor cholera was recorded in India, at Calcutta, and by October of the same year it had become the principal aetiologic agent of cholera in that town.

The origin of the disease in Calcutta was open to debate; it could not have derived from Bangladesh in view of the lysotype but had probably been imported from another country by a sea route. Others assumed that it was not imported but rather a variant modified by the chlorine added to the town water (the laboratory data likewise appeared to sustain this point of view).

The appearance of El Tor was not an isolated case since in the course of an epidemic in Gujarat both *Vibrio cholerae* and *Vibrio El Tor* were isolated.

The westward trend of cholera became more evident when it penetrated into Bombay in 1964, then passed through Pakistan to Afghanistan (1965), Iran (July 1965) and Iraq (1966).

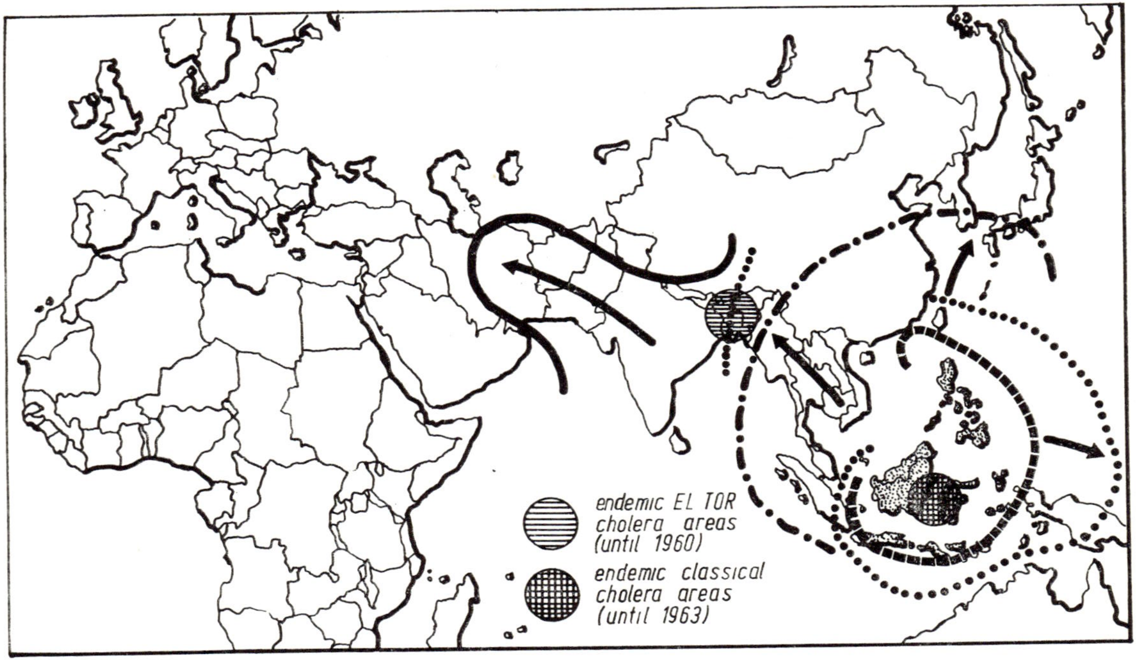

The propagation of cholera (1961–1965) (WHO Chronicle, 1970, 20, 7)
FIG. 5.2.

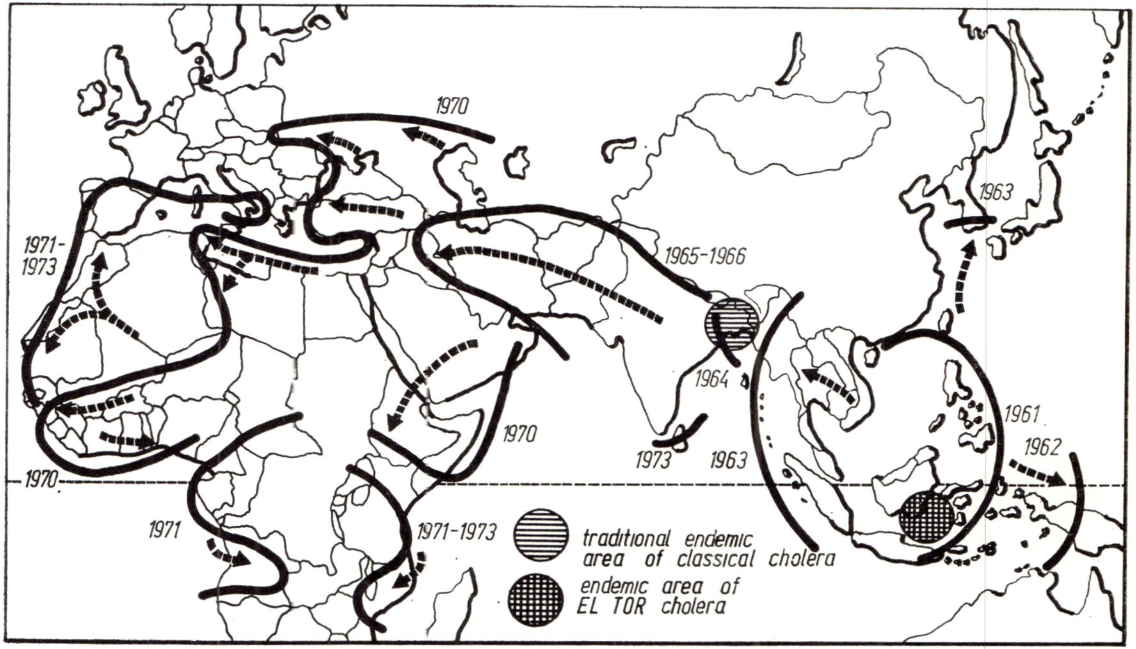

Global spread of cholera 1961—1973

FIG. 5.3.

This focus in the Near East (Iraq) took on an extensive character in 1970 following the events in the neighbourhood of the Suez Canal, where troops were massed. Cholera gradually spread to Israel, Jordan, Turkey, the U.S.S.R.

However, spread of the epidemic depended upon the health organisations of the infected countries; the consequences were minor in the countries with good health organisations but progressed alarmingly in Africa, along three main lines.

(1) In the north, from Egypt it penetrated into Syria, Tunisia, Morocco and Algeria, the countries of the Maghreb becoming an endemo-epidemic focus.

(2) To the east, from Egypt it spread to Somalia, Ethiopia (where an important cholera focus developed), Sudan, Kenya, Uganda, Ruanda and Tanzania.

(3) To the west, in less than a year, cholera struck at almost all the western and central African countries, with a starting point in Guinea (August 1970), along the coast to Cameroon (March, 1971), then to the countries of the Sahel, starting with Mali (November, 1970), then along the Niger, penetrating into Upper Volta, Niger, Nigeria, North Cameroon (May, 1971) and finally in Chad, Central Africa (June, 1971), Mauritania and East Senegal (summer of 1971).

In December 1970, Felix wrote: "the epidemic in Africa is taking on a disastrous aspect and we do not see how it can be controlled. What is now occurring in Mali and the Ivory Coast, what is beginning in Dahomey, sets us thinking that cholera has come to stay for many years on the western coast of Africa".

On the western coast cholera had the classical aspect of an epidemic with a dominant hydric component, whereas at first along the Niger then in the dry desert regions the disease spread by direct interhuman communication, starting from patients, cadavers and carriers, promiscuity and sweat playing — in the opinion of Felix (1971) — a major role. The same authors states that the number of cholera cases was more than 100 000 with 20 000 deaths, the case fatality rate being much greater in the central zone.

In the Sahel region, cholera started in the cold, rainy season and spread along the course of the rivers borne by native craft, and in the dry hot season by lorries, public transport and the roads through the grassland — markets and feasts were key places where one patient or a few carriers triggered an epidemic.

Behind the epidemic wave, the disease did not die out, persisting in the form of local endemics with epidemic recrudescence in the rainy season.

At first the El Tor vibrio, Ogawa type was isolated, but towards the end of the African epidemic the Inaba and Hikojima serotypes were isolated.

In March 1971 there was a severe epidemic recrudescence in the traditional focus of Bangladesh, extending along the Gulf of Bengal. This last pandemic again brought up the seriousness of the problem of cholera, and raised the question of whether the El Tor vibrio would pass the frontiers of Europe.

Already, from the beginning of the seventh pandemic, cholera manifested a tendency to penetrate into the European countries. The evolution of the imported cases differed: in some countries there were only a few primary cases, in others, patients discovered late gave rise to secondary cases. In 1970, in Czechoslovakia and France there were only primary cases, in the Soviet Union and Turkey there were also secondary cases (see Table 5.1).

TABLE 5. 1.

Cholera in European Countries (1970, WHO)

(i) imported cases; (ii) epidemics

Country	No. of cases	Isolated biotype and serotype	Infection circumstances
U.S.S.R ii)	92 Odessa 87 Kerch	El Tor Ogawa	First cases in Astrakhan
Czechoslovakia i)	4	El Tor Ogawa	3 tourists
France i)	1	V. cholerae Ogawa	Tourist in North Africa
Turkey ii)	4/1181	El Tor Inaba	

In 1971 in Europe, the number of cholera cases was small but the number of imported focuses increased (Table 5.2).

Except for one case from which the El Tor Hikojima serotype was isolated, all the others were El Tor Ogawa serotype. In Portugal

TABLE 5. 2.

Cholera in European Countries (1971, WHO)

(i) imported cases; (ii) epidemics

Country	No. of cases	Isolated biotype and serotype	Infection circumstances
France[i]	3	(1) El Tor Hikojima (2) El Tor Ogawa	(1) imported from Cameroon (2) tourists from Barcelona
Portugal[ii]	4	El Tor Ogawa	imported
Spain[ii]	22	El Tor Ogawa	tourists Morocco
Sweden[i]	3	El Tor Ogawa	imported from Spain
West Berlin[i]	1	El Tor Ogawa	vacation in Spain
England[i]	3	El Tor Ogawa	imported Barcelona-Malaga

and Spain secondary cases appeared; in the other countries there were only primary cases and the focus was considered extinct. The cases in the other countries were imported by tourists going or coming from Spain, or the African countries.

In 1972 there is an evident regress of the imported cases and focuses, giving a false impression that cholera no longer represented a real danger for the developed European countries. The WHO considered that imported cholera did not represent a danger for the countries with a high health status and well trained medical corps. However, this point of view was not confirmed by the subsequent situation. In 1973, the number of imported cases in European countries reached a peak at the beginning of the seventh pandemic (Table 5.3; Fig. 5.4).

TABLE 5. 3.

Cholera in European Countries (1973, WHO)

Country	No. of cases / month	Isolated biotype and serotype	Infection circumstances
England	1/April	—	—
France	1/April	—	Excursion to the Ivory Coast
England	3/Aug.	El Tor Ogawa	Tourist from Tunisia
West Germany	2/Aug.		Tourist from Tunisia
West Germany	1/Aug.	El Tor Ogawa	Tourist from Tunisia
Sweden	10/Aug.	El Tor Ogawa	Tourist from Tunisia
France	2/Aug.	El Tor Ogawa	North Africa
Italy	10/220/Aug.-Sept.	El Tor Ogawa	North Africa
West Germany	2/Sept.	—	Tourist from Tunisia
West Germany	1/Sept.	El Tor Ogawa	Excursion to Italy

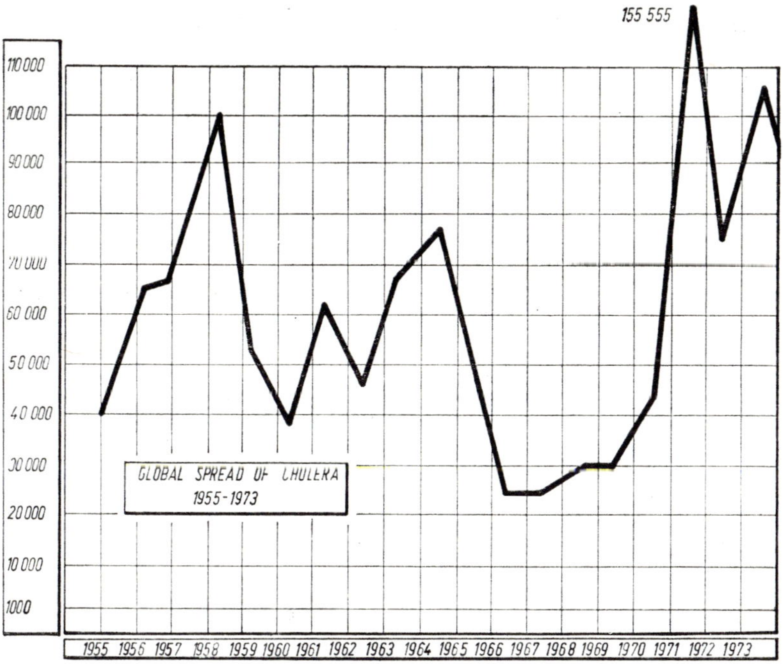

Cholera, morbidity during the 1955—1973 period

FIG. 5.4.

From the data of the WHO published up to October 1 1973, El Tor cholera was imported into West Germany, five times, four times into England, twice into France and once into Sweden and Italy (where the primary cases precipitated a focus of 150 cases in the Bari region). From all the cases the El Tor Ogawa serotype was isolated except for one imported case in West Germany (1 Sept. 1973) from which the El Tor Inaba type was isolated. As a rule the imported cases were tourists from the respective countries who had been on an excursion to North Africa (Tunisia).

These data suggest, we believe, that cholera still represents a real and a potential danger. Comparison of the first six to the seventh pandemic (Table 5.4) reveals some of the particular traits of the present pandemic in which the El Tor biotype plays the main role.

It is worth underlining the properties of the El Tor biotype, which make it more redoutable than the classical *V. cholerae:*

TABLE 5. 4.

Characteristics of the seventh pandemic in relation to the first six ones

Characteristics	First six pandemics	Seventh pandemic
Origin	Indo-Pakistan focus	Indonesian focus
Etiologic agent	V. cholerae biotype	El Torbio type more resitant in the environment
A tendency to spread	From east to west	Only westward
The course of the pandemic	Return to the initial focus after a given circuit	Tendency to become established in the newly invaded areas
Carrier status	No carriers	Frequent, even longstanding carriers
Clinical forms	Classical, sometimes benign	Mostly of enterocolitic type
Therapeutic possibilities	Not well developed High case fatality rate	Therapy well established. Case fatality rate 1 %.
Prevention (vaccination)	Doubtful results (sometimes good) (Romanian experience in the Balkan war)	Vaccination not satisfactory

(i) El Tor is better endowed and adapts better to the environment than the classical strain;

(ii) the classical cholera used to pass periodically beyond the borderlines of its endemic focus but returned comparatively soon to its starting point. In contrast, El Tor cholera shows a tendency to pass beyond the geographical limits of the endemic zones and to become established in the areas in which it has penetrated (see Africa);

(iii) El Tor causes mild forms that are often mistaken for mild enterocolitis in the summer season (commonly encountered in most import-foci in Europe). Cholera becomes suspect when the typical cases appear, and then it is too late;

(iv) worthy of note is the appearance of El Tor strains resistant to tetracyclin, which demands a certain reserve in the use of this antibiotic as a mass preventive.

The control of cholera and prevention of its import must therefore be adapted to the new situation.

This outline of the history of the El Tor vibrio with its great penetrative potential brings to the fore the actual danger of cholera and its spread to intact countries and shows that:

(i) cholera has been on the offensive since 1961;

(ii) the starting point is now in Asia, but not at this time in India or Indonesia;

(iii) the prospects are of a new, vast endemic zone in Africa, representing a permanent danger for Europe;

(iv) the causal agent is no longer the classical *Vibrio comma* but the El Tor biotype.

5. 2. Physiopathogenic mechanism

Both the classical and El Tor vibrio penetrates into the body by oral route, the only route of infection. This has been proved experimentally; parenteral administration (intramuscular, intraperitoneal) failed in the rabbit (Dutta *et al.*, 1959—1960), the infective dose being evaluated at 10^6 germs.

In the stomach the vibrio encounters an important natural barrier, the acidity, which can destroy it. In the absence of acidity (in the aged, patients suffering from anacidity, extensive resection of the stomach, etc.) or low acidity the vibrio may pass into the intestine whose alkaline pH favours its survival and multiplication. In the lumen of the small intestine (jejunum) the vibrios multiply and release several toxic fractions, the most important being enterotoxin which appears to be responsible for the severe intestinal and humoral disturbances characteristic of cholera. Also called *choleragen*, this enterotoxin brings about a massive, sudden passage of water and electrolytes from the vessels of the intestinal wall towards the intestinal lumen, followed by marked elimination of fluid and electrolytes in the stools.

Up to 15—20 litres may be lost. The fluid part of the blood is massively eliminated producing rapid haemoconcentration and marked increase in the viscosity of the blood. This depletion also results in accentuated loss of electrolytes (Na, K, Cl) and especially of the alkaline substances required for acid-base balance, followed by acidosis (marked loss of potassium through the stools, in excess of the chlorine lost by vomiting).

The loss of potassium leads to nervous and muscular disorders, intense cramps with muscular exhaustion (metabolic paralysis), extrasystoles and cardiac muscular alterations (evident on the ECG trace) and aggravation of circulatory disturbances. This aspect is more conspicuous in the stage of secondary collapse.

Blood calcium falls, manifested clinically by tetany. Urea and non-protein nitrate increase. Severe cases are accompanied by an increase in bilirubinaemia (indirect fraction) attributed to the intravascular haemolytic action of the cholera toxin.

The toxic coordinates of cholera have not been fully established. It is not yet understood whether the loss of water and electrolytes depends upon the local action of the enterotoxin or on general toxaemia; this is contested by most authors.

Opinions concerning the intimate mechanism of the loss of water and electrolytes still remain a moot point.

Most specialists consider that there is no inflammation at the level of the intestinal mucosa, which is not damaged.

Others sustain the presence of lesions in the intestinal wall similar to those encountered in the malabsorption syndrome or idiopathic steatorhoea. As argument they invoke the presence of C reactive protein even when biopsy of the intestinal mucosa is negative. According to Sack *et al* (1970) the presence of reactive C protein is a more convincing argument for local inflammation than simple, specific interference with passage of the fluids through the intestinal wall, as sustained by others.

Normally, in a man weighing 60—70 kg there is a passage of fluid from the plasma towards the intestinal lumen and inversely of

80—100 litres in 24 hours. According to Philips (1967) two mechanisms contribute separately or together to the massive evacuations in cholera: (1) an irresistible discharge of water and electrolytes without proteins from the blood towards the intestine; (2) a decrease in the transfer of water and electrolytes from the intestine towards the blood. Wisscher (1944) showed that the sodium ion brings about this passage through the intestinal wall. This mechanism was suggestively dubbed by Phillips the "sodium pump". Irrespective of their opinion most agree that in cholera inhibition of the "sodium pump" occurs.

From the beginning this inhibitor was considered a toxic fraction (Watten and Phillips, 1960; Fuhrman and Fuhrman, 1960). The presence of the inhibitor was demonstrated in the stool filtrate of cholera patients. The inhibitor proved to be thermolabile.

Dutta and Oza (1963) isolated from cholera vibrio acid extracts a choleragenic substance which, administered to laboratory animals (young rabbits for instance), produces diarrhoea and death. It may be assumed that these inhibitors are secreted in the intestine and disintegration of the vibrios facilitates their release, acting both locally and in the entire body.

The choleragenic substances may be destroyed by enzymes of the phosphorylase type extracted from muscles, but are not influenced by amylase or lipase.

Starting from these findings Dutta and Oza (1963) considered that the pathogenic activity of the cholera toxin lies in its lipid and polysaccharide fractions.

Enzymes of the phosphorylase type only act in older rabbits which prove resistant to infection by intestinal route; young rabbits show a marked sensitivity, contracting the disease even after a small infective dose (10^4).

In 1965, Craig isolated from a vibrio culture filtrate a "permeability factor" that acts directly upon the vascular endothelium bringing about an increase in permeability. Intradermally, this toxic fraction produces prolonged erythema and oedema (confirming the increased permeability of the small cutaneous blood vessels). The permeability factor is considered to be identical to enterotoxin (diarrhoeogenic factor). Lending support to this assumption is also the fact that the administration of antitoxin neutralises the intracutaneous vascular effect of this factor.

Recent works have made clear the pathogenic mechanism to a great extent. Diarrhoea is caused by the cholera vibrio enterotoxin.

For all types of cholera vibrios this enterotoxin is a protein with a molecular weight of 84 000, sensitive to heat and an acid medium resistant to trypsin.

In the cells of the intestinal epithelium there are two enzymes: adenylcyclase, which mediates the action of the enterotoxin, and adenosine cyclic monophosphate which, under certain conditions, prevents the absorption of sodium.

After release of the enterotoxin in the small intestine it becomes rapidly and irreversibly bound to the epithelium, so that the administration of large amounts of antitoxin two minutes after the administration of toxin does not influence the loss of fluid induced by enterotoxin.

The enterotoxin acts as a hormone, stimulating the activity of adenylcyclase which causes the accumulation of adenosine cyclic monophosphate in the epithelial cells of the intestine. This arrests the passage of sodium through the intestinal lumen towards the intestinal capillaries and favours the inverse flow that draws after it the respective amount of water. This mechanism or "sodium pump of the intestine" (Phillips, 1967) is responsible for the hydroelectrolytic disorders of cholera.

The massive loss of fluids and electrolytes induce circulatory failure and collapse up to shock, with haemoconcentration, oliguria and anuria.

The lesions in the liver, kidneys and other organs are secondary to hypoxia, the direct consequence of peripheral circulatory insufficiency.

In the pathogeny of cholera other organic factors interfere. In childhood, old age, pregnancy and denutrition, people are more receptive and the forms of the disease more severe.

The course of the disease may also be unfavourably influenced by other concomitant intestinal infections or parasitoses, frequently, encountered in tropical countries and warm climates. This explains some of the intestinal lesions found at necropsy.

The pathologic physiology of water and electrolyte depletion are not yet sufficiently clear to be able to find a vaccine acting at the level of the intestine.

5. 3. Pathologic anatomy and pathogenetic relationships

The first histologic data were collected at necropsy; autopsy of the fatal cases revealed epithelial desquamation of the small intestine.

It has been assumed that enteritis in cholera is manifested by three independent characteristic processes: inflammation, repair and degeneration.

The intensity of the inflammatory process is not accentuated and the reparatory process is manifested by compensatory hyperplasia with numerous mitoses in Lieberkühn's crypts. The degenerative process is believed to affect especially the epithelium covering the extremity of the villi, running parallel to the formation of lacunae in the basement membrane, as a precursor sign of subepithelial oedema, and detached epithelium, found at autopsy. The atrophic lesions of chronic enteritis may sometimes also be observed in the more distal cells of the epithelium.

Intestinal biopsy demonstrated, however, that these are actually artefacts occurring after death or existing during lifetime in the case of a concurrent parasitic or bacterial disease.

In the course of cholera the epithelium of the intestinal villi remains histologically intact. Congestion of the mucosa alone takes place due to the vascular effect of the cholera toxin (permeability factor) that accounts for the reddish tint of the mucosa at necropsy (hortensia-like aspect).

The fundamental lesions of the intestine in cholera affect the vessels of the wall, resulting in degeneration of the vascular endothelium (the election action of the permeability factor). In common diarrhoeas the lesions differ from those found in cholera only by their lesser intensity.

In all the cases a considerable production of mucus was found in the Lieberkühn glands.

In conclusion, the intestinal mucosa remains intact during the patient's lifetime and the lesions classically described are due to autolysis produced by digestive enzymes and/or by *post-mortem* bacterial invasion. On the other hand, the rice-water stools considered as proof of vital detachment of the intestinal mucosa, does not exceed the normal renewal of a healthy mucosa; the rice-water aspect is caused by the abnormal production of mucus.

6. Clinical picture

Incubation in cholera is very short — on average three days — but may vary from one to seven days.

The 3 to 5 days incubation period was adopted by the WHO in the International Health Regulations (1951).

In general, the shorter the incubation the more severe the disease is. However, fatal cases have also been seen following an incubation of 5 to 6 days.

As the clinical picture of cholera is unique, and in the classical cholera epidemics the clinical symptoms present various degrees of intensity, we shall give a description of the typical form of *cholera gravis*, irrespective of whether it is caused by the classical *Vibrio* or *Vibrio El Tor*.

6. 1. Evacuation stage

The onset is sudden, in full health, with diarrhoeic stools, dominating the clinical picture and characterising the first stage of evacuation.

Within the first hours there are still traces of faeces but subsequently the stools become watery, effortless, without cramps, increasing in number and amount.

Massive amounts of fluid are discharged, 5 to 6 litres/24 hours without any control of the anal sphincter.

The second characteristic sympton of cholera, encountered in 85% of the cases, is vomiting, concomitant to or even preceding diarrhoea (as a rule a few hours after the onset of diarrhoea). Irrespective of the moment in which it appears, vomiting is very dangerous, not so much because of the loss of water it entails (which is not great) or electrolytes (chlorine especially) but because it prevents oral administration of antibiotics and rehydration by this route.

If these two main symptoms cannot be controlled they inexorably end in death.

According to El-Ramli (1948), the duration of diarrhoea is 12 hours to 15 days (on average 4 to 6 days), and of vomiting 1 to 8 days (on average, 3 days).

Vomiting as a rule soon stops and it becomes possible to administer antibiotics orally and to take up progressive rehydration.

The evolution of the stools is almost always characteristic. After they have become fluid and turbid with no traces of bile, whitish, rice-like grains appear and the stools take on the characteristic rice-water aspect.

The grains contain leucocytes, droplets of fat and/or cells of the internal layer lining the villi and tubular glands of the intestine; part of the stools are transformed into an amorphous mucoid mass.

Cholera vibrios may also be found in variable amounts, sometimes in pure culture.

According to the number of red blood cells the stools may take on a pinkish colour.

The vomit consists at first of the contents of the stomach with the rest of food, then is transformed into an uncharacteristic fluid that becomes rice-waterish similar to the stools.

Another characteristic symptom of the onset period are the muscular cramps encountered in 60–70% of the cases. Initially localised in the calves, later they may appear at any site: the muscles of the fingers, hands, thighs, face. The cramps are ushered in by a sensation of stretching and "pins and needles", then set in sharply and last a few minutes. On palpation the muscle is hard and rigid, and becomes all the more so with progressive dehydration and loss of the elasticity of the tissues. The cramps appear to be stronger in young, vigorous individuals and run parallel with the loss of water and electrolytes, especially potassium.

These three elements — diarrhoea, vomiting and muscular cramps — form the symptomatic triad of the onset phase of cholera, preparing for the subsequent drama that may occur within 2 to 3 days.

The clinical picture of the evacuation stage is very clear cut. In contrast to variola, especially the haemorrhagic form, the cholera patient is quiet, apathetic. He exhibits a considerable degree of anorexia and a permanent thirst, but any attempt at drinking is followed by vomiting.

As dehydration becomes more accentuated and in terms of the rate at which it develops the patient takes on a characteristic aspect: dry skin tending to form folds that no longer stretch out; the abdomen begins to fall in. The face is likewise characteristic, with dark circles round the eyes, prominent nose, dry and purplish lips.

Although most authors include these disorders in the evacuation stage, Desphande (Kasturba Hospital – Bombay) divides this period into two stages: a period of evacuation and one of dehydration, they follow one another, the latter being the consequence of the former.

Gradually, new signs develop reflecting severe intoxication of the entire economy of the body:

(i) the blood pressure falls, the systolic pressure falling below 90 and even 60 mmHg;
(ii) the pulse rate increases slightly but does not exceed 100 min^{-1};
(iii) the body temperature is at first normal then gradually falls;
(iv) the amount of urine decreases;
(v) hiccoughing appears and becomes very difficult to bear.

All these signs portend aggravation and transition to the next stage, that of collapse.

6. 2. The stage of collapse

The transition to this stage is continuous but may be very sudden (within a few hours) in the severe forms.

With time new signs of gravity develop:

(i) circulatory insufficiency (blood pressure falls below 50 mmHg);
(ii) tachycardia (120 beats min^{-1}), the pulse becomes thread-like and imperceptible;
(iii) the heart sounds are dull and weak;
(iv) the skin and extremities are cold;
(v) anuria.

In this collapse stage, the initial symptoms – diarrhoea, vomiting and cramps – disappear as a rule, but may continue and contribute still further to deterioration of the clinical picture.

Dehydration reaches a peak in this stage of algid collapse, and the patient is in an advanced stage of emaciation, so that one can truly say he is only "skin and bone". The skin is cold and moist, the lips and nails cyanotic; the patient lies back in bed, his eyes sunken with an anxious look on his face; his tongue is dry and fuliginous.

El-Ramli also mentions oppression of the chest and, at the clinical examination pericardiac friction and an imperceptible pulse with zero blood pressure.

To sum up: circulatory failure is maximum in the stage of algid collapse. Initially of peripheral origin—due to the massive extraction of fluids from the vascular bed, emptying of the vessels and increase in the concentration and viscosity of the blood – collapse subsequently takes on a central component due to involvement of the heart. The cardiac sounds become weaker and the second sound disappears completely; systolic murmurs appear. It is considered that, apart from the direct action of the loss of water and electrolytes (especially potassium, which interferes in contraction of the myocardium) the cholera toxin exercises a direct action on the cardiac fibre.

The deficient circulation and hyperconcentration of the blood mass lowers the force of the cardiac contraction due to the poor nutritional supply; the ECG tracing points to acute coronary failure.

In some cases the patient complains of retrosternal pains and a galloping sound can be heard at the base of the heart.

Heart failure is one of the causes of death; a few hours before dying the patient presents bursts of extrasystoles, the precursors of ventricular fibrillation.

Renal failure is always present, accompanied by circulatory insufficiency and, as a direct outcome, renal anoxia with decrease of the glomerular filtrate, oliguria and anuria.

The Indian authors Das Gupta *et al.* (1967) assume the existence of a renal shunt that blocks completely the supply of blood to the glomeruli.

The danger of uraemia increases with the continuation of collapse.

In general, renal and circulatory failure go hand in hand and are the main cause of death in the stage of algid collapse.

Respiratory insufficiency is seldom encountered; during the most recent epidemics dyspnoea has not been described nor the sensation of pressure on the chest caused by disturbance of pulmonary oxygenation, consequent to circulatory disorders. Pleural and pericardiac friction can be heard.

Dysphonia is characterised by loss of voice; the patient in his extreme weakness can no longer articulate, his vocal cords are dry; accordingly, total aphonia is a sign of extreme gravity.

The sensory organs are also affected. The patient's face takes on a fixed expression, with glassy, dry eyes and slow movements of the pupils. Dark brown spots sometimes surround the dilated vessels of the conjunctiva. A tendency to opacification of the cornea may be observed. Miosis has been considered as characteristic of the algid stage; in the terminal phase of collapse the pupils are dilated.

Alterations are rare in the other organs, except the ear.

The body temperature has been much discussed. At first the temperature is normal or even low towards the extremities with a constant tendency to continue falling; a temperature of 30 °C or even less (but not the buccal temperature) is characteristic of the collapse stage.

In contrast to the extremely low peripheral temperature in some cases with a fatal end, rectal temperature may rise to 39 – 40 °C a few hours before death.

The duration of the algid stage represents the turning point in the evolution and prognosis of the disease: the shorter it is the more chances the patient has of survival, the longer the fewer chances.

This stage is now, however, rarely encountered thanks to the emergency treatment applied. When the treatment is not started in time, the course of the disease progresses with extreme coldness of the extremities, cold viscous perspiration, half-open eyes and fixed stare, arrest of the cardiac activity, stertorous respiration and death.

Algid collapse lasts from a few hours to 1–2 days; when it lasts more than 8–10 hours the chances of recovery are very small.

6. 3. The reaction stage

Today, due to a correctly applied emergency treatment for replacing water and electrolyte losses, the development of cholera may be arrested before reaching the irreversible stage of decompensated hypovolaemic shock that causes death.

Recovery of the patient has been described by several authors and, in general, their description agrees.

Accordingly, the rapid reaction without complications takes place as follows:

(i) the pulse begins to beat slowly and irregularly, then becomes stronger and more regular;

(ii) breathing is easier and the patient calmer;

(iii) cyanosis and chill, as well as viscous perspiration of the skin gradually disappears and the temperature returns to normal or even a little higher;

(iv) the stomach recovers its tonus and the food is no longer thrown up;

(v) the faeces are coloured (sometimes with mild diarrhoea, slightly watery stools, greyish brown in colour);

(iv) urine is passed, at first in small concentrated amounts; the smell is strong, the density high, with albumin, indican and casts, then gradually becomes more abundant;

(vii) the patient gradually recovers his physical and muscular strength.

The process is not continuous but irregular and frequently interrupted by relapses. This occurs especially in cases that show signs of recovery from collapse but in which progress is slow, irregular and incomplete. The body temperature rises but the extremities still remain cold. Urine may continue to be passed in small amounts. A certain degree of circulatory insufficiency persists and in such cases death may occur suddenly in spite of the apparent tendency to recovery.

The patient is exhausted and relapses to a state of algid collapse may occur once or several times. These complications are chiefly due to renal failure or uraemia.

Finally, remission may be accompanied by two exaggerated reactions: hyperthermia and uraemia.

There is no general consensus regarding hyperthermia. Apart from that caused by pyrogenic substances in the saline perfusion fluid, Rogers (1952) noticed a frequent increase in the body temperature of European patients who died after having survived the phase of algid collapse — some of whom had not received saline perfusions (28% of the deaths during the reaction period).

The duration of the reaction period depends upon the presence or absence of an excessive response of the body and especially upon the appearance of complications.

In cases without complications the patient recovers completely after 5 to 7 days although slight diarrhoea may persist for a week.

The danger of a relapse, however, may threaten the patient within the first 7 to 10 days of convalescence. These relapses are often caused by overeating, the use of laxatives, premature exposure to the cold or physical effort, etc.

When the relapse occurs much later (after more than two weeks) the question of a reinfection may be raised.

6. 4. Clinical forms

We have described in the foregoing passages the classical clinical form of cholera as encountered by many physicians in the course of cholera epidemics, the El Tor epidemic included. We have tried to emphasise the gravity and permanent threat of dehydration and algid collapse.

To underrate the danger of cholera caused by *Vibrio El Tor* is a serious mistake since:

(i) certain forms of El Tor cholera may be just as severe as cholera gravis;

(ii) the evolution of cholera cannot be foreseen and when not treated immediately and correctly a mild form may become severe and end in death.

The pathogenic potential of the El Tor vibrio has not yet been precisely established; the epidemics of Celebes included many severe forms whereas those of Calcutta, Nepal and Philippines had a high proportion of mild cases.

In their order of gravity, the clinical forms of cholera are:

(1) *The fulminating clinical form — cholera siderans* — also called *cholera sicca* is the most severe form and 100% lethal; a single case is known to have been saved. However, this form has seldom been encountered during the more recent epidemics.

The disease begins in full health, without diarrhoea, with signs of severe intoxication manifested by cyanosis, hyperthermia and collapse. The patient dies before being able to evacuate the rice-water material in his intestine.

(2) *Cholera gravis* is accompanied by algid collapse and was the form usually encountered in the first six waves of cholera in the last century, being considered as a standard of gravity.

This form was also encountered in El Tor epidemics especially the first epidemic in Celebes in 1958 when a case fatality rate of 90% was registered.

(3) The *moderate clinical form* of cholera is characterised by an onset with the characteristic symptoms of the disease: diarrhoea, vomiting and muscular cramps. In this form the loss of water and electrolytes is not so great and the patients do not reach the stage of algid collapse; they may pass through an intermediate stage of precollapse in which therapy with saline perfusions re-equilibrates the body. This form of the disease is chiefly encountered in El Tor epidemics. However, lack of treatment or delayed treatment will aggravate the clinical picture which may become that of cholera gravis.

These moderate forms were encountered in the epidemics of:

(i) Surabaya in which more than half of the 79 fatal cases evolved as moderate forms up to the moment of their aggravation;

(ii) Philippines in which two thirds of the deaths (13% mortality rate) were treated late or not at all;

(iii) the U.S.S.R., Turkey, Iraq where a high proportion of moderate forms was met.

Characteristic of this form is the evacuation stage that attests to cholera intoxication, manifested by:

(i) numerous, spontaneous rice-water stools;

(ii) vomiting without nausea for 2—3 days;

(iii) muscular cramps.

This form is therefore characterised by a variable degree of dehydration, moderate collapse or its absence.

(4) *Choleriform gastroenteritis* is the most frequent clinical form encountered: next to the abortive and subclinical forms it is considered as classical of El Tor cholera. Described in the Malaya epidemic, this clinical form has the features of acute gastroenteritis. The five mainsymptoms are:

(i) diarrhoea encountered in 99 % of patients has a sudden onset, with abdominal pains and lasts 1 to 2 days. In two thirds of the patients the number of stools ranges between 6 to 12 a day. About 40% of the patients have typical rice-water stools;

(ii) vomiting, present in two thirds of the cases, concomitant with diarrhoea of moderate intensity;

(iii) abdominal colics in a quarter of the patients with diffuse pains but without abdominal defence;

(iv) fever 37—38°C in about 20% of the cases;

(v) muscular cramps.

The onset of these symptoms, irrespective of their intensity, is always sudden and the diagnosis readily established in case of an epidemic. In the Malaya epidemic 530 adults and 186 children were admitted to hospital; 205 of the total number of cases were confirmed clinically and 128 also bacteriologically. They considered that the other non-diagnosed cases were also cholera.

Attesting to cholera intoxication are the symptoms mentioned: diarrhoea, vomiting without nausea and muscular cramps. It should be stressed that dehydration is not very accentuated and the stage of algid collapse does not appear; the moderate collapse or rather precollapse, depends upon the therapy (when not treated the stage of algic collapse may readily develop).

(5) *The abortive or attenuated forms* start with the characteristic symptoms of cholera but have neither the duration nor the intensity of the other forms.

Particular clinical forms are encountered in the aged and in children, in whom they are the more severe the younger the child (Lindenbaum *et al.*, 1966).

In the neonate, the disease is 100% lethal, evolving rapidly, even without diarrhoea, with toxic collapse, agitation, muscular cramps, hypothermia and anuria.

In the aged, diarrhoea represents an immense danger; the disease evolves with asthenia, marked hypothermia and total physical exhaustion, causing death in 50—70% of the cases.

(6) *The typhoid form of cholera* described by the classical authors, has been the subject of many discussions and is characterised by a typhoid state, with high fever within the context of the characteristic symptoms of cholera.

(7) *The subclinical forms* may be found among healthy carriers and appear to be proper to the infection with El Tor vibrio. Healthy carriers are asymptomatic. After digestive disturbances clinical signs may appear.

From the foregoing data it will be seen that the clinical picture of cholera is the same regardless of whether it is caused by the classical *Vibrio* or *Vibrio El Tor.*

The following deductions may be drawn:

(1) Cholera remains a severe disease, with an unforeseeable potential and evolution; the moderate or mild forms may readily be transformed into severe cases.

(2) El Tor cholera exhibits an evident potential for causing epidemics; its clinical forms cover the entire range of classical cholera and even if, at present, the moderate and mild forms are prevalent, we do not know what the future holds for us.

6. 5. Complications

Complications are inherent to cholera — a disease that creates a deep imbalance in the economy of the body, affecting the circulatory, renal, respiratory, digestive sytems, etc. The complications may develop in the course of the disease and during convalescence. In view of the extreme gravity of cholera, the complications may pass unobserved during the disease, the major symptoms being predominant.

Apart from the disorders caused by the cholera toxin, which may persist with various consequences, suprainfections with pyogens may still be met with today in various tissues and organs deeply affected by the disease.

The *gastrointestinal tract* often remains irritable and with a tendency to diarrhoea long after the acute attack. This is seen especially in patients with a past history of digestive defects, tropical malaria, bacterial or amoebic dysentery, metabolic and denutrition disturbances, etc.

In such cases permanent anorexia develops apart from the persistent, tenacious diarrhoea.

The liver, bile ducts and gallbladder may become the site of acute bacterial complications. The gallbladder may become an occasional or permanent host and play an important role in the carrier status. Bile carriers are intermittent excreters and with each acute attack excrete considerable amounts of vibrios.

The *circulatory system.* A variable degree of bradycardia may persist during convalescence and even long afterwards. The toxin acts on the myocardium with severe consequences, and a new infectious disease develops that affects the heart.

The *respiratory tract* is the site of disturbances that may continue for some time: bronchitis, pneumonia, bronchopneumonia which are always a severe threat. Pleural exudates and suppurations are less often encountered.

Today pulmonary bacterial complications and those occurring at the level of other organs are more readily treated with antibiotics, obviously after having overcome the acute dehydration stage.

Pulmonary oedema, which appears in the reaction stage is in many instances the outcome of untimely administration of fluids exceeding the limits necessary for hydration, and stressing an already deficient heart. Pulmonary oedema may also be accompanied by uraemia.

The *urinary tract* bears with difficulty the handicap of a circulatory insufficiency. The more acute and rapid is the renal failure the greater are the chances that the kidney will be left with severe lesions.

Suprainfections in various segments of the urinary tract may be favoured by local conditions.

The *skin* is the site of multiple complications in the course of the disease, more frequently in the past than today. Carbuncles, abscesses, phlegmons, eschars may appear when hygiene is deficient.

Seldom before, and altogether exceptional today, is the occurrence of meningitis, ulcers of the cornea, otitis.

The pre-existing diseases or those contracted during convalescence play an important part in aggravating the disease (especially bacterial or amoebic dysentery). Pregnancy is a severe hazard; the outcome is abortion, premature birth or stillbirths and/or death of the mother.

6. 6. Clinical and epidemiological diagnosis

The most important and difficult problem is the diagnosis of the first cholera case.

A correct and rapid diagnosis of the first case must be based upon clinical, epidemiological and laboratory data.

The clinical and epidemiological diagnosis involves detailed survey of all travellers coming from a cholera focus and presenting signs of acute digestive disturbances. The following need to be established:

(i) period during which the traveller stayed in the respective cholera region;
(ii) contact with cholera patients or convalescents;
(iii) moment in which he left the cholera focus;
(iv) when the first symptoms appeared.

As cholera is an acute disease with a short period of incubation these details must be established precisely.

Any digestive disease with diarrhoea, with or without rice-water stools with a sudden onset, marked dehydration, intoxication and a tendency to collapse, must be considered as suspect.

The clinical and epidemiological data must be completed by laboratory findings even if the patient does not exhibit all the characteristic symptoms of the onset of cholera.

Under the present epidemiological conditions cholera caused by the El Tor vibrio comes foremost, with its prevalent atypical clinical forms, with or without cholera-like features, such as mild enterocolitis without fever. Neither should the possibility of a subclinical form be overlooked. The carriers may present at different intervals acute digestive syndromes that are considered an alimentary imbalance or intercurrent digestive infections. In such cases the longer incubation period, exceeding the known limit, may render the diagnosis more difficult; their past history may be marked by a previous acute digestive period.

Former patients who have become biliary carriers may be decompensated periodically in the course of an inflammatory attack of the bile ducts or gallbladder (cholangitis, cholecystitis). The excretion of vibrios make of them an important epidemiologic source.

6. 7. Differential diagnosis

In countries with endemic cholera the differential diagnosis consists first in establishing the causal agent: the classical cholera or the El Tor vibrio. For intact countries this is not significant since the clinical picture is the same. This is also the point of view of the WHO which recommends the same measures in both cases.

In intact countries the differential diagnosis is made with alimentary toxic infections since *Vibrio El Tor* frequently generates a picture of acute enterocolitis with or without cholera-like aspects. In contrast however to cholera, which evolves without fever and with muscular cramps, in alimentary toxic infections fever is almost always present as well as abdominal colics but without muscular cramps. The differential diagnosis should also be made with various intoxications of a vegetable, animal or chemical nature, and with infantile diarrhoea *(pseudocholera infantum)*.

Napier (1951) systematised the differential diagnosis between cholera, alimentary toxic infections and intoxication with arsenic and antimony.

In special circumstances the differential diagnosis is established with postantibiotic choleriform staphylococcal enterocolitis. In this syndrome there is a severe loss of water and electrolytes that set in brusquely after prolonged treatment with antibiotics.

In this severe complication there may be rice-water stools but the intestinal content is usually purulent, odourless, and effortless; anamnesis is of particular importance in such cases.

Intestinal anthrax may give rise to confusion, as it is characterised by sanguinolent enterocolitis with deep alteration of the patient's general condition. The past history and especially the presence of *Bacillus anthracis* in the faeces elucidates the diagnosis.

Bacterial dysentery differs from cholera by its specific mucosanguinolent stools, abdominal tenesmus, without muscular cramps, with fever and slightly altered general status, without collapse. The presence of the causal agent in the faeces and the characteristic endoscopic aspect, with sanguinolent ulcerations of the mucosa and erosive sphincteritis establish the diagnosis.

In amoebic dysentery there is a moderate loss of water without alteration of the patient's general condition. The "button-like" endoscopic lesions and the presence of the dysentery amoeba are important diagnostic indications.

The association or reawakening of a chronic dysentery after cholera are not excluded.

Malaria with digestive manifestations imposes a differential diagnosis. The local digestive symptoms in acute malaria, dominated by profuse watery stools, associated with collapse and a pernicious algid stage, are factors that raise difficulties when establishing a diagnosis of cholera. Examination of the smears can clarify the diagnosis.

Trichinosis in the digestive stage may give rise to confusion but urticaria and eosinophilia differentiate it from cholera.

In tropical countries intestinal giardiasis may sometime sinduce a false interpretation, especially in cases with massive infection and violent diarrhoea. The presence of the parasite gives the diagnosis.

Various vegetable, animal and mineral toxic substances may provoke clinical pictures similar to cholera. The presence of the toxic substance and anamnesis are the decisive diagnostic factors.

7. Management of cholera

7. 1. Treatment

The exceptionally large amounts of fluid lost by cholera patients in the stools had been observed in ancient times. Substitution of these losses was recommended for the first time in 1831. But electrolytes are also lost, being withdrawn from the blood by the discharge of fluids. The attempts to use water alone as a substitute were not successful; on the contrary, they aggravated the situation, especially in the severe cases in which the natural route of hydration cannot be used.

Rogers (1915) has the merit of having introduced parenteral administration of fluids, using hypertonic saline solutions with which he substantially improved treatment of the severe forms of cholera.

In spite of initial criticisms, hypertonic solutions have gradually been adopted by all the physicians called upon to look after cholera patients.

According to the data of Chun (see Pollitzer, 1959), culled in the course of two epidemics, the mortality rate was as in Table 7.1.

TABLE 7. 1.

	1919	1926
Mortality rate of patients treated with hypertonic solutions	14.11%	17.2%
Mortality rate of patients treated with isotonic solutions	54.5%	34.5%

At that time, however, little was known of the physiopathogenic mechanism oft he disease, nor about the composition of the stools in cholera; progress was made only after detailed analysis of cholera stools.

Phillips initiated the first studies in this direction; he showed that in cholera the intestinal epithelium maintains its integrity. This was also proved by Gangarosa *et al.* (1960) with a Crosby capsule and subsequently confirmed by intestinal biopsy.

Analysis of the stools led Phillips (1963) to conclude that they are isotonic, closely resembling the blood plasma.

The following electrolytes are lost in the faeces:

Sodium	150 mEq l^{-1}
Potassium	10–15 mEq l^{-1}
Chlorine	110 mEq l^{-1}
Bicarbonates	40 mEq l^{-1}

Children lose somewhat more sodium but less chlorine.

Phillips proposed to replace these losses by isotonic solutions closely resembling the blood. Hypertonic saline solutions were in favour in India and are still used in some places today. In Bombay, Desphande began rehydration of cholera patients with hypertonic saline solution (500 ml) administered at a rapid rate.

These hypertonic solutions were not, however, fully satisfactory. Although they produced a rapid effect when administered in small amounts, they caused thirst and agitation caused and the short term hydroelectrolytic balance obtained was considered unsatisfactory by many doctors, which explains why the traditional intravenous alkaline saline treatment proposed by Rogers has undergone so many changes.

In the Hospital of Communicable Diseases of Calcutta the following standard solution is used:

(i) hypertonic saline solution (1.35%) (600 ml);
(ii) isotonic saline solution (600 ml);
(iii) 5% dextrose solution in normal saline (600 ml);
(iv) 7.5% sodium bicarbonate solution (50 ml);

The treatment is applied immediately after admission and repeated after 3 hours in patients with a weak or imperceptable pulse. Special investigations into this treatment were carried out in the Autumn of 1963 and Spring of 1964 in order to establish the causes of death in patients treated with this solution and the comparative clinical development of confirmed cholera patients and those with non-cholera acute diarrhoea.

Isotonic solutions began to be used increasingly, and several injectable solutions for correcting dehydration and the loss of electrolytes were proposed:

(i) normal saline;

(ii) sodium bicarbonate isotonic solution (14% in distilled water);

(iii) 1/6 M sodium lactate solution sodium lactate (18.7 g) in water 1000 ml;

(iv) 2% sodium bicarbonate solution;

(v) 5% glucose solution in normal saline (1000 ml) to which sodium chloride (10 mg) is added;

(vi) Ringer solution;

(vii) Dacca solution (5-4-1) sodium chloride (5 g) sodium bicarbonate (4 g) and potassium chloride (1 g) in distilled water (1000 ml). With these solutions combined in different ways it is initially possible to reduce dehydration, then to regain and maintain the hydroelectrolytic balance (Greenough *et al.*, 1964).

For correct effective treatment the following must be taken into account:

(i) the degree of dehydration and electrolyte losses that have to be determined;

(ii) choice of the solutions to be administered;

(iii) the rate of administration and the amounts to be administered in stages;

(iv) the total amount of fluids necessary;

(v) criteria for the control of a correct hydration.

The *degree of dehydration* may be determined by measuring the fluids lost in the stools, urine and vomit, and clinically by the arterial pressure and pulse.

The following accurate methods may be used:

(i) determination of the plasma volume with Evans blue;

(ii) determination of radioactive chromium labelled erythrocytes;

(iii) measurement of the central venous pressure;

(iv) determination of the plasma specific weight (normal 1024—1027) with copper sulphate.

The total amounts of fluid used in the course of rehydration (indirectly measuring the degree of dehydration and the volume of fluid lost) may be calculated within the first 2—3 hours by determining the initial plasma density with copper sulphate; for each increase in density of 0.001 above the normal (1.025) the required fluid volume is of 4 ml kg^{-1} body weight.

The wide range of solutions offers many possibilities. In general the choice is based upon the results obtained in analysis of the composition of the stools (electrolytes and bases).

Normal saline frequently combined with alkaline solutions (sodium bicarbonate or lactate) and potassium or glucose may be added if necessary.

Good results have also been obtained with the Dacca solution (5-4-1).

Carpenter *et al.* (1966 a) recommend the following scheme: normal saline + Na lactate at first in a proportion of 2:1 then 1:2. A similar scheme was applied in Madras and Calcutta, with, however, a modified rate of administration.

The intravenous route of administration has been selected, especially for the first hours of treatment when rapid intervention is vital for remission from hypovolaemic shock and acidosis.

Intravenous administration of these solutions may be followed by untoward effects: chills generated by the pyrogenicity of the solutions, irritation at the site of the injection because of the catheters, or along the veins, in case of prolonged perfusions, etc. Moreover, in epidemics large volumes of solutions are necessary and the stocks may soon be depleted. The administration of these solutions demands numerous highly, skilled personnel. In children, the intraperitoneal route is preferred.

In 1962, the first attempts at oral rehydration were made (Phillips, 1963). It was found that the cholera patient can absorb water, potassium and bicarbonates administered orally but not sodium and chlorine. Glucose, however, may be administered and with it sodium and chlorine ions. These data led to reconsideration of the oral route.

One of the solutions used contains:

(i) sodium	120 mEq l^{-1}
(ii) chlorine	72 mEq l^{-1}
(iii) bicarbonates	48 mEq l^{-1}
(iv) sodium citrate	15 mEq l^{-1}
(v) glucose	110 millimoles l^{-1}

in heated drinking water to facilitate dissolution. The solution is kept in 15 l plastic bags (Cash and Nalin, 1970).

Another solution giving similar results contains rather less potassium and sodium — 10 mEq and 110 mEq l^{-1} (Pierce et al., 1970 c).

The solution is administered either directly orally or by nasal catheter so that the patient may be able to rest.

When the patient is in shock the treatment must start with intravenous perfusion and continue until normal blood pressure, pulse rate and plasma density (below 1027) is obtained; from then on hydration may be continued by oral route.

In the mild cases with a blood pressure of 100 mmHg and a perceptible pulse, hydration may be started orally from the beginning.

The rate of administration and the necessary amounts are established in terms of the patient's condition and degree of dehydration.

Patients in shock need a more rapid rate of administration to fill the vascular bed as soon as possible and to correct acidosis (manifested by dyspnoea of the Kussmaul type), the loss of electrolytes — hypopotassaemia (manifested by cramps and muscular exhaustion, arrhythmia, ballooning).

In the hospital at Bombay the first 500 ml hypertonic saline is administered in 1 to 2 minutes. In the severe cases the rate of administration may be 100 ml min^{-1}.

In the epidemic of Malacca were introduced 500 ml in 5 minutes after which the rate of perfusion was slowed down. The same rate is applied in the hospital at Madras where 1 litre is initially introduced in 10 minutes (the solution is distributed in 500 ml plastic bags). It is considered that 40% of the patients need about 4 litres. In the initial period an adult patient is given about 2.5—3 litres.

The order of administration is the following:

(i) N saline solution (0.5 litre);

(ii) N saline solution (0.5 litre);

(iii) chloride solution (0.5 litre) with 2% sodium bicarbonate 7.5% sodium bicarbonate (150 ml) in N saline (440 ml) or 1/6 M sodium lactate (0.5 litre);

(iv) N saline (0.5 litre);

(v) N saline (0.5 litre).

The amounts and rate of administration differ however with age:

for children aged 8—12 years	2.5—3 l
4—8	1.5—2 l
1—4	0.5—1.5 l
0—1	125 375 ml

As a rule the patient recovers from collapse after 20 minutes with about 2.5 litres solution; if this fails then the perfusion must be repeated after 4 hours, administering at most 15 litres in 36 hours.

In the hospital at Poona (Gracias, 1967) 50 ml are perfused per minute, then as the patient's condition improves 25 ml min^{-1} until the blood pressure and pulse return to normal. On average a patient receives 20 litres.

In the Calcutta hospital the rate of administration is much slower: 3 to 4 litres in 3 to 4 hours. In general, 10 litres per patient are considered sufficient and only in exceptional cases are 25 to 40 litres administered.

The saline solution packed in plastic bags and kept in the cold is apyrogenic and sterile and may be stored indefinitely. The first stage of rehydration is over after 4 to 5 hours.

In a second stage, tetracyclin is added and then in a third stage potassium salts (coconut milk).

Based upon their experience the Indian physicians generally administer the necessary amounts of fluid without laboratory tests (in epidemics the number of patients demands an enormous contribution on the part of the laboratory with which the personnel cannot always cope).

As a rule the veins in the bend of the arm are used for perfusion.

In order to ensure a high rate of perfusion (1 litre in 10 to 15 minutes) there must be two perfusion points, and thick needles have to be used. The perfusion flasks must be suspended at least 2 m above the bed. If the veins of the arms are not patent the subclavicular or the femoral vein, and in children the external jugular vein, may be used. As the situation improves perfusion will be continued at selected sites.

In cardiac patients saline solutions must be administered with care; similar precautions must be taken on administering potassium solutions to renal patients.

Particular attention must be paid to the rate and amount of fluids administered to children in order to avoid hyperhydration.

In children a smaller amount of sodium and a larger amount of potassium is lost in the stools, and the irhydric balance is more critical.

It is recommended that Ringer's solution with sodium lactate which also contains potassium be used. Rehydration should be done carefully in order to avoid an exaggerated, brutal supply of Na. The duration of perfusion is 8 hours, half the amount being administered within the first 2 to 3 hours. The patient must be constantly monitored in order to avoid overloading with fluids.

The administration of fluids is in itself a problem. The veins currently used in the adult become impracticable, and the external jugular veins must be used and less often the epicranial vein which demands a certain skill.

For practical reasons some authors resort to the intraperitoneal route, which does not, however, permit rapid resorption, nor does it give the expected results in severe cases of dehydration (Mahalanabis *et al.*, 1970).

The fluid intake is gradually increased according to the patient's digestive tolerance, since it may produce or exaggerate vomiting. About 600—1500 ml are given per hour. More than 1500 ml is contra-indicated.

The amount is calculated in terms of the fluid loss during the last four hours.

Rehydration may be continued orally. In this case, per os administration may be applied 3 to 4 hours after the beginning of the intravenous treatment, when the pulse and the arterial blood pressure, as well as the specific plasma weight, become normal again.

In most cases oral administration of fluids maintains the equilibrium and only in cases of refractory vomiting and numerous stools is it necessary to take up intravenous perfusion again and supplement it by oral administration.

The oral intake of fluids solves many problems. The injectable fluid requirements may be reduced by 70—80% which is very important in epidemics since the solutions administered by oral route are prepared from drinking water. They need not be sterilised, for the ingredients are cheap and readily available, advantages which in epidemics, especially in rural areas, are obvious (Nalin *et al.*, 1968 and 1971; Nalin and Cahs, 1970; Sack *et al.*, 1970).

The sooner the patient is admitted to hospital and the intravenous treatment started the sooner will the oral treatment begin, favourably influencing the clinical evolution and prognosis of the disease.

Once the initial water and electrolyte losses are replaced, the current losses in the course of the disease have to be compensated, hence the losses both in stools and vomit have to be measured. As long as the patient vomits compensatory fluids will be supplied by the intravenous route, using one of the solutions mentioned (Ringer + lactate, the 5-4-1 solution, etc.). Once digestive tolerance is re-established rehydration may be continued by oral route.

In this way, the total fluid intake calculated at 7% of the patient's body weight (according to the Soviet authors) and up to the body weight (Phillips, 1967) may be considerably reduced in the course of the treatment.

The total amount of fluid necessary depends upon: the gravity of the case and the moment in which the treatment was instituted — both the symptomatic treatment for replacement of water and electrolytes and antibiotic therapy.

In an adult, hydric losses are about 4—5 litre, after which he may continue to lose in the stools and vomit about 0.25 litres. An efficient antibiotic treatment will reduce these secondary losses by half. Fluid requirements in a case of medium intensity are of up to 7 to 8 litres in the adult and about a third of this amount in the mild forms.

According to the WHO experts the criteria of a correct, efficient rehydration are:

(i) normal blood pressure and pulse;

(ii) recovery of cutaneous turgor;

(iii) good general condition made evident by disappearance of cyanosis, cramps and vomiting;

(iv) normal diuresis after 12—24 hours treatment;

(v) return to normal of the specific plasma density (below 1027).

Incorrectly applied treatment is made evident by:

(i) fever (chills) either due to the injectable pyrogenic solutions or to use of the same rubber tube for several flasks;

(ii) persistance or reappearance of dehydration with hypovolaemia-acidosis-shock; insufficient amounts of fluid or premature interruption of rehydration;

(iii) continued nausea and vomiting—insufficient supply of bases;

(iv) tetany — due to excessive supply of water and alkali;

(v) acute circulatory failure due to an insufficient supply of water;

(vi) renal insufficiency due to prolonged hypotension (insufficient replacement of the lost fluids);

(vii) hypopotassaemia which in children may precipitate arhythmia, asthenia, abdominal distension; convulsions, coma in children due to excessive supply of water and sodium;

(viii) hyperhydration, marked by dilatation of the veins of the neck and slow pulse, which often precede a characteristic heart failure, especially in children and adolescents;

(ix) hypoglycemia, frequent in children, due to an insufficient supply of glucose;

(x) acute pulmonary oedema caused by excess fluids introduced intravenously.

In conclusion, the pathogenic treatment is the fundamental element in the emergency therapy of the severe forms of cholera; thanks to this the proportion of recoveries may be as much as 99.%

Today the pathogenic treatment is completed (in the second stage) by an anti-infectious therapy with antibiotics and chemotherapeutical substances which are very efficient in clinical and bacteriologic recovery (Carpenter *et al.*, 1964 and 1966 b; Lindenbaum *et al.*, 1967 a and b; Uylancgo *et al.*, 1967 b; Wallace *et al.*, 1968). The administration of these drugs begins as soon as the patient has recovered from shock.

The moment at which the antibiotic can be introduced may be hastened by a rapid and correct pathogenetic treatment. Administration runs parallel to digestive tolerance by the oral route. The antibiotic is chosen in terms of the sensitivity spectrum (established by the antibiogram).

The cholera vibrio is sensitive to a wide range of antibiotics: tetracyclin, chloramphenicol, erythromycin, novobiomycin, neomycin, streptomycin, paramomycin, gramicidin, cycloserin, etc. as well as to chemotherapeutical agents, especially bactrin (septrin) and nitrofuran derivatives (furazolidin, furalazin), sulfonamides, nalidixic acid, etc.

Stricytol (streptomycin + thalazol) associations have likewise been used in Malacca.

The best results have been obtained with tetracyclin, considered as the optional antibiotic; it has a definitely favourable influence on the course of the disease when administered in doses of 2 g per 24 h for 3 – 5 days. Carpenter (1969) obtained excellent results after 3 days, with a rapid decrease of water depletion, as compared to the control sample not treated with antibiotics:

(i) on day 2 the antibiotic treated sample lost about 2 200 g fluids as against 4 300 g in the control sample;

(ii) on day 3 those treated with antibiotics lost 490 g as against 990 g in the control sample.

In the antibiotics sample the cholera vibrio disappeared from the stools after 24 hours whereas in the control sample elimination lasted on average 6 days. Similar results were reported by Greenough *et al.*, 1964.

It is important that the treatment should last at least 3 days in order to obtain sterilisation of the bacteriologic relapses.

In the treatment of cholera its duration is more important than the amount of antibiotic administered.

The most efficient route of administration is the oral one, with direct action at the level of the body-bacteria conflict. The intravenous route recommended by some authors appears to be less active in view of the pathogenic mechanism of cholera.

Consequent to the use and abuse of antibiotics certain strains resistant to one or several antibiotics, tetracyclin included, have made their appearance (Philippines, Turkey, U.S.S.R.). In certain situations it is recommended to use an antibiotic to which the vibrio proves sensitive – chloramphenicol, erythromycin, or still better bactrin-furazolidin – or combined antibiotic and chemotherapeutic.

The treatment with choleraphage, formerly recommended and then abandoned, has now been taken up again, but the investigations are limited and must be checked in larger samples.

A comparative study has shown that the administration to children for five days of small tetracyclin doses (100 mg day^{-1}) and furazolidin gives results similar to those obtained only with tetracyclin (500 mg day^{-1}) (Pierce *et al.*, 1968).

The treatment is completed by diet and hygiene: rest in bed, alimentation as soon as digestive tolerance is reestablished and the patient asks for food, adding fruit juices, rice soup, lemonade, coffee, mineral waters, coconut milk, etc.

Vitamins C, B, etc., analeptics, sedatives – in case of convulsions – corticoids for controlling toxaemia, antihistaminics, etc. may be added.

Today the treatment of cholera arrests the disease which formerly gave such a high case fatality rate in the untreated cases – up to 60%.

The present simple treatment has demonstrated its efficiency in the last pandemic in which the case fatality rate fell to 1%.

The last pandemic evidently raised once again the problem of improving the treatment, based upon new data concerning the pathogenesis of cholera, digestive toxic infections causing massive loss of water and electrolytes by the intestinal and oral route.

To the symptomatic, pathogenic and anti-infectious treatment an antitoxic therapy may be foreseen in the future. This involves stimulation of the adenylcyclase system – cyclic AMP, an effect comparable to that exercised by certain prostaglandins – directly responsible for water and electrolyte losses.

It has been possible to inhibit, experimentally, the outpouring of water and electrolytes by various substances (cyclohexamide, ethacrinic acid, acetazolamide).

In the mouse, oral administration of some anti-inflammatory drugs (salicylates, steroids, pyrazols) have the ability to reduce water and electrolyte losses caused by cholera toxins. These substances will not, however, be able to replace rehydration, which is the priority medical intervention, both simple and efficient.

To conclude, here are some practical indications of an orientative character:

(i) the patients must be admitted to hospital as soon as possible;

(ii) the degree of dehydration must be determined by measuring:

(a) blood pressure, pulse, turgor, etc.;

(b) the fluids lost must be measured;

(c) laboratory tests (specific plasma weight);

(iii) rehydration and correction of electrolyte losses;

(iv) choice of the perfusion solution, preferably normal saline or an alkaline solution (with sodium bicarbonate or lactate) to which potassium may be added if necessary;

(v) obligatory administration by the intravenous route;

(vi) rate of administration of up to 100 ml min^{-1} of a total amount of 2–2.5 litres;

(vii) clinical follow up of the patients and determination of the laboratory constants. From the moment in which the pulse and blood pressure return to normal intravenous perfusion will be replaced by oral administration of the solutions which must contain glucose;

(viii) the addition to the treatment of antibiotics (more efficient by the oral route after vomiting has ceased);

(ix) continued rehydration by oral route until diarrhoea ceases; the diet is completed by vegetables, soup, compote to which little sugar is added, coffee, lemonade, puree, sauté, etc.).

7. 2. Prognosis

The gravity of the disease is reflected in the high mortality rate which in the untreated cases is still very high today in some regions (40–80%).

The pathogenic treatment lowers the mortality rate in adults to 1% or even less. Similar results were obtained in children.

In the extremities of the age groups (infants and old people) the disease is still severe in spite of the treatment; pregnant women are likewise severely affected (Hirschhorn *et al.,* 1969).

The number of severe cases is, however, smaller in general in the epidemics caused by El Tor vibrio. In the severe cases following a rapid, correct treatment, death is an exception.

However, even today the prognosis is not always certain and in some cases the situation may suddenly worsen. In order to prevent an unfavourable course the patient must be admitted as early as possible and carefully examined for any signs of gravity: typhoid state, massive dehydration with collapse, intense muscular cramps, thoracic oppression, etc.

A sharp fall in temperature (below normal values), prolonged algid collapse, persistent anuria, pulmonary, renal, hepatic complications are signs of aggravation that lower the patient's chances of recovery.

An early competent treatment transforms cholera into a disease that can be cured.

7. 3. The cholera hospital

In countries with endemo-epidemics cholera patients are admitted to special wards or in annexes to communicable diseases hospitals.

In the inter-epidemic periods all the patients detected may be admitted but during an epidemic episode this is no longer possible and only patients with massive water and electrolyte losses whose life is in danger are admitted to hospital. In some countries, for example India, it is not compulsory to bring cholera patients to hospital.

Although eradicated in Europe owing to improved sanitary conditions, cholera still remains a permanent threat especially in the new epidemiologic conditions created by *Vibrio El Tor*. In Europe the appearance of a cholera case is an unusual epidemiologic event.

The regulations foresee in general two distinct stages:

(i) measures to be taken up to the appearance of the first case(s);

(ii) and after the first cholera cases have appeared, in peacetime and in case of social calamities (earthquake, floods, etc.).

For the prevention and control of cholera the health officials must establish:

(i) the hospital where cholera patients will be taken;

(ii) the institute (reference laboratory) which will confirm the diagnosis of cholera by bacteriologic tests, for which it is fully responsible. Any error of diagnosis may have important implications.

The prevention and control of communicable diseases in general and of those subject to quarantine in particular is based upon the complex cooperation between the health inspectorate, hospital and laboratory.

As already emphasised for smallpox, this activity must be constant and operative, involving information and close communication, which at a given moment will trigger the ensemble of the measures to be taken in case of cholera.

7. 3. 1. The hospital

Hospitals in the capital city of the country and of the various districts which will take in cholera patients must be established previously. A hospital for communicable diseases where specialists are to be found is chosen as a rule. Other hospitals may, however, also be chosen since in an epidemic the number of cases may be very large.

It is recommended that a hospital built on the pavilion system be chosen not a building with several floors. The hospital should be situated far from water sources (underground water layer, running water, etc.) and the drinking water treatment units of the town. The water pipes and sewage system of the hospital must be in perfect order.

The water and sewage systems must not cross one another nor come in contact with those of the neighbouring buildings. Before joining the main sewers, the hospital residual waters must be treated. In no instance must the sewage be discharged directly into surface waters. The hospital water system (W.C., baths, sinks, etc.) must be periodically checked also in the cholera pavilion.

There must be no exposed refuse dump favouring the multiplication of flies.

The hospital must be separated from its neighbourhood by a fence. As regards the access and inner distribution of the compartments in the bungalow the rules are the same as for the smallpox hospital (at least two entries, special cubicles for the patients, food larder, clean and dirty linen kept separately, storeroom for disinfectants and general utilities, sterilisation facilities, drugs and instruments, etc.).

Particular attention must be paid to the routes followed for dirty linen, food, etc., as well as anything that might favour transmission of the disease by water, bedclothes, linen, food.

It is not necessary to provide accommodation for members of the staff since they may leave the pavilion after taking the necessary measures of disinfection. The state of health of the personnel has to be carefully checked.

The pavilions are for admission and treatment of confirmed cholera patients, and each one should be separated into two distinct parts:

(a) a section reserved for resuscitation of the patients in shock whose condition demands intensive therapy;

(b) a section for milder cases, to which the patients that have recovered from shock are moved. If possible the suspects will also be admitted and kept in a separate ward until the diagnosis is confirmed bacteriologically or invalidated, when they are discharged.

Special provision should be made for the quarantine of contacts (coming from endemic regions).

The annexes for the isolation and care of cholera patients have to provide for:

(i) isolation of the patients up to complete clinical and bacteriological recovery;

(ii) specialised treatment, particularly for the severe cases;

(iii) bacteriological diagnosis (from the pathologic products of the patients);

(iv) diagnosis of the carrier status;

(v) sterilisation of the carriers;

(vi) quarantine of persons coming from endemo-epidemic regions.

The staff of the pavilions should include 1 or 2 doctors, a head (assistant) nurse, four medical assistants, two nurses, a dietician, two ambulance men, two porters.

The experience of the Ivory Coast epidemic showed that one nurse or medical assistant was necessary for ten patients during an epidemic with a continuous flow of patients.

The medical and nursing staff should be vaccinated against cholera. Chemoprophylaxis or antibiotic therapy are useless; as long as the rules of individual and collective hygiene and prevention are complied with regarding manipulation of the pathologic products (stools, vomit, linen, etc.) the risks of contracting the disease are minimal.

The medical and ancillary staff must attend special training courses to be able to cope with severe cases suffering from accentuated dehydration (collapse-shock), particular attention being given to the administration of perfusions. For this, each member of the staff must be acquainted with:

(i) the perfusion solutions;

(ii) preparation and handling of the equipment (kits, flasks, etc.);

(iii) the technique of administration (rate);

(iv) risks (drawbacks, complications, especially chills) and their prevention and treatment;

(v) the measures to be taken concerning the patient's pathologic products.

Refresher courses will also include the treatment and management of severe alimentary toxic infections and other diseases with an accentuated loss of fluids and electrolytes.

The courses should be complete with documentary films, booklets, lectures and practical demonstrations so that the nurses will also know how emergency treatment is applied in cases of collapse even in the absence of the doctor.

The hospital should be supplied with:

(i) beds with an opening in which a graduated vessel is introduced for the stools and urine discharged so as to measure the amount of water lost which has to be replaced by perfusion (Fig. 7.1). These

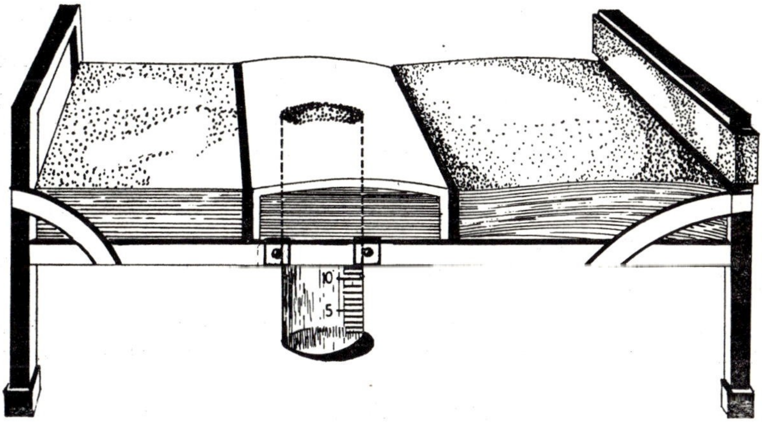

Special bed for cholera patients, for use in hospital. Note the central opening and graduated container for collecting and measuring the fluid lost so as to calculate substitution requirements.

FIG. 7.1.

special beds must be provided for the intensive therapy sections; in the other wards simple hospital beds may be used. The beds are provided on both sides with metal stands for suspending the perfusion flasks;

(ii) mackintosh and drawsheets;

(iii) at least 6 bed linen sets for each bed;

(iv) graduated vessels for collecting the stools and urine;

(v) urinals, bedpans, wash basins, sputum mugs, preferably of plastic that can be thrown away;

(vi) wire netting at the windows to prevent the flies from entering;

As cholera is transmitted by hydric route, the bathroom and laundry are particularly important links in the cholera annexe.

The water supply and sewage system must be controlled. It is important that there should be no leakage and that all the joints should be carefully sealed so that there is no danger of the water being contaminated by sewage.

The bath must be disinfected each time it is used, as well as the floor and walls of the bathroom. No dirty linen or other materials may be kept in the bathroom.

There will be a separate laundry for the cholera unit with separate circuits for the clean and the dirty clothes. The dirty linen will be brought to the laundry in special bags after being disinfected. When a separate laundry is not available a military type mobile laundry can be annexed to the bungalow.

When sewers do not exist, in rural areas for instance, the bath and laundry residual waters will be collected in draining wells and constantly disinfected. However, the system is not 100% safe and cannot be considered as a substitute for a treatment unit.

The pharmacy must be prepared to put at the disposal of the cholera unit the required drugs and perfusion solutions (normal saline, glucose, bicarbonate, etc.). The perfusion solutions may be stored in plastic bags (as used in the Hospital for Communicable Diseases in Calcutta).

In order to be able to cover the requirements, especially the large amounts of perfusion solutions needed, the pharmacy must be provided with:

(i) a high capacity still for bidistilled or distilled water;

(ii) chemicals (pure, *pro injectione)* for preparing the solutions;

(iii) the reserve solutions must be gradually replaced so as not to exceed the storage limit;

(iv) the necessary chemicals for preparing the oral solutions (glucose, NaCl, KCl, etc.).

The fever-inducing property of the solutions prepared must be periodically controlled as chills represent a serious danger for cholera patients in collapse.

The clinical laboratory plays an important role in the correct management of the therapy. There are numerous laboratory criteria for appraising the degree of dehydration which vary from case to case and in terms of local possibilities and the volume of work required. Today controlled, efficient re-equilibration may be obtained, thanks to these tests.

The most frequently used are: plasma density, ionogram, alkaline reserve, etc.

7. 3. 2. Measures taken on admission of the patient

As soon as a cholera case is notified, the doctor on duty must open the envelope containing the instructions regarding cholera.

He must call up the director of the hospital and send a car to fetch him. If the director is not in town he will call up the senior medical officer assigned to the cholera unit and then the members of the staff after checking their names and address.

Once the team has arrived they will be vaccinated against cholera according to the vaccination scheme. The cases with temporary contra-indications will be given either fractional doses or replaced.

The unit or annexe assigned to cholera patients will then be evacuated without creating panic to other sections or even hospitals. This should be done stepwise, one unit after the other in order of emergency.

The wards are rapidly inspected, the sanitary installation, service pipes, etc. controlled, access to the unit closed and the equipment completed:

(i) beds, night tables, mattresses, pillows, sheets, cupboards for the linen;

(ii) food larder, cupboards for the dishes and cutlery;

(iii) various objects: graduated vessels for the bed, urinals, bedpans, wash basins, bags for the dirty linen;

(iv) disinfectants for the disinfection unit;

(v) instruments and kits;

(vi) cupboards for instruments.

The patients' circuit is studied; as far as possible the patient should be brought to the cubicle where he will remain until recovery. The hospital entrance will be closed, and a poster erected with the words "QUARANTINE. ENTRANCE FORBIDDEN". No other patient may enter from this moment. The doctor on duty will decide by which gate the ambulance should enter (not the usual one) and communicate this to the inspectorate which coordinates evacuation of the patient. He will then put somebody at the gate to show the ambulance driver the way to the cholera unit. In no case must he stop at the admision consultation room.

From the moment the patient is admitted to the unit the protection perimeter (fence) is closed and nobody allowed to enter except with a special permit signed by the hospital director.

At the entrance to the unit a sign "CHOLERA" will be put up.

The pathologic products collected during transfer to hospital (stools, vomit) will be taken in the respective vessels to the laboratory, under all necessary precautions.

The ambulance must be disinfected before leaving the hospital precincts and the driver must have a written paper attesting to the disinfection.

All the personnel that accompanied or came in contact with the patient or his belongings are considered contacts and must be vaccinated and kept under medical survey. At the first digestive disturbances he should be isolated until the diagnosis is established.

After the cholera patient has been admitted to the annexe the hospital entrance is opened again for other patients.

The activities within the cholera unit are completely separate from those in the rest of the hospital.

If the patient's condition permits he is taken to the bathroom and undressed. All his clothes and linen are introduced into a container with a disinfectant. He is then washed with soap and water, given hospital linen and put to bed. His case card is completed, noting the date and hour at which he was brought in.

Parallel to this examination, which may be more detailed or complete according to the patient's general condition and degree of dehydration, intravenous perfusions are started to correct losses of water and electrolytes. Before beginning the treatment, however, the first pathological products are collected (stools, vomit), packed, labelled and sent to the laboratory. Blood is withdrawn for determining the loss of water and electrolytes and checking the rehydration treatment.

Not for a moment should one forget that the patient's life is in danger and that each minute counts. The patient's life depends upon the rapidity with which the treatment is instituted and the hydroionic balance corrected.

7. 3. 3. Management of the cholera hospital

The data presented are orientative the same as for smallpox. They may be modified and improved from case to case, according to local conditions, but with strict compliance to the various circuits established for transit of personnel, pathological material, etc.

If the risk of contamination with smallpox is very great, dust from the patient's ward representing the chief source of infection, in cholera, transmission by the digestive route is very important (water, food, flies), and the fact that explosive epidemic outbreaks may occur in centres with a common water supply should not be overlooked, especially when the population is not resistant to the infection. Active immunisation is compulsory in such cases but cannot be considered to confer complete safety.

In the cholera building compliance with the circuits referred to above is a fundamental necessity.

The first circuit is that for the patients. As already mentioned, the patient should be brought directly to the cubicle where he will remain until complete recovery. The door through which he has entered is then closed.

The cubicles with beds should, if possible, be arranged along a central corridor.

The doctor's visits to patients should be made according to a certain routine. At the entrance to the first cubicle there must be a small recess with hangers for the protective clothes (bonnet, blouse, trousers, slip covers for putting over the shoes).

In each cubicle there must be soap and water and disinfectant for members of the staff to wash their hands after examining or nursing the patient. After the last cubicle there is another recess in which the personnel take off the protective clothes and put them in containers with an antiseptic solution from which they are taken to the disinfection unit. A mask for the mouth and nose and spectacles are not necessary.

After changing, the personnel will wash their hands with soap and water or take a shower bath. This circuit ends at the disinfection unit.

Any member of the staff who enters the patient's cubicle must proceed in the same way.

The food circuit is similar to that described for the smallpox pavilion. The food is prepared separately in the hospital kitchen and brought to the cholera unit through a door specially assigned for this purpose. On a mat soaked in an antiseptic solution are the kitchen containers for the annexe. The hospital orderly puts the food into the containers without touching them and after he has left the dietician takes the food from the containers and serves the patients.

The dishes and cutlery are washed and introduced in an antiseptic bath, then rinsed.

The remains of the food are burnt in an incinerator.

The clean linen circuit starts at the clean linen cupboard at a point opposite to the disinfection unit, and ends in the patient's cubicle. The dirty linen circuit begins in the patient's cubicle and ends in the disinfection unit.

As with smallpox, dirty linen is carried in plastic bags steeped in an antiseptic solution.

Both the drugs and the instruments must have similar circuits.

Particular attention must be paid to the pathologic products, especially faeces and vomit. The stools gathered in the graduated vessels below the patients' beds (the amount excreted is measured) or in bedpans or other containers are disinfected then flushed down the water closet. Disinfectant is absolutely necessary no matter how safe the state of the sewage system is considered to be or the fact that there is a treatment plant.

Thoroughness is one of the main points upon which one must insist on when chosing the personnel.

If part of the stools or pathologic products should fall on the floor they must be covered with an antiseptic solution, left for some time and then removed.

The cubicles must be kept spotlessly clean, and the floors wiped with specially impregnated mops kept in an antiseptic bath.

Patients remain in hospital until they have recovered completely, and after three negative coprologic examinations they may be discharged.

If the vibrio still persists in the stools the treatment with tetracyclin is resumed and three control coprocultures are performed.

8. The laboratory diagnosis of cholera

Both in the endemic cholera regions and in the intact countries threatened by the spread of a pandemic, laboratory tests play an outstanding role in the control of cholera.

The cholera laboratory tests done as early as possible are absolutely necessary in view of:

(i) the urgency of the measures to be taken in case of a cholera focus;

(ii) the importance of a diagnosis for elucidating certain clinical forms (cholera, cholera infection or vibrio-free cholera — as described by the WHO experts) and identification of the non-cholera vibrios;

(iii) detection of the carrier status, more frequent and prolonged today and caused especially by the El Tor vibrio.

Identification of all infected persons and of the routes of transmission are based upon a thorough knowledge of the pathogenic agents isolated and the accuracy of the work of the laboratories assigned to the diagnosis of cholera within a certain area.

Bearing this in mind the present chapter will refer to the properties of the *V. cholerae* and the laboratory diagnosis of cholera.

8. 1. Vibrio cholerae

Discovered in 1883 by Koch the vibrio of cholera forms part of the *Vibrio* genus *(Vibrio comma, Vibrio cholerae)* in which vibrios with different characters, including those non-pathogenic for mankind, are included.

At present the cholera vibrio includes both the classical vibrio which has been very well defined and characterised, and the El Tor vibrio.

With the spread of cholera, starting in 1961, the pathogenicity of El Tor was fully recognised; as its properties are very similar to the classical strain it was considered to be a biotype of the same species, with similar epidemiological and clinical implications. According to the WHO both may be considered as cholera vibrios and in the following pages we shall refer to it in this accepted use of the term.

Hugh (1965) proposed the nomenclature of *Vibrio cholerae* and *Vibrio cholerae El Tor* biotype; when necessary we shall mention the type, i. e. classical vibrio and El Tor vibrio.

8. 1. 1. Characteristics of *Vibrio cholerae* (Table 8.1)

After a rapid review of the well-known data in text-books we shall emphasize the characteristics that are of importance for our understanding of the present manifestations of cholera, its treatment and control, and particularly the necessity for a rapid, correct diagnosis.

Morphology: the cholera vibrio may present, apart from the characteristic aspect (slightly curved rods 1—5 μm long), certain variations that appear in terms of the product examined (samples from various levels of the intestine; laboratory cultures on solid or fluid media; environmental elements — water, soil, food, etc.), period of examination (beginning or end of the disease; beginning or end of the epidemic; age of the culture); the influence of antibiotics, disinfectants, bacteriophages. The various forms range from granular to long or even filamentous.

Worthy of note is the mobility of the vibrios, the largest of the bacteria, that facilitates their identification in fresh products. Cholera vibrios swarm like fish shoals. Antibiotic treatment reduces their mobility.

The cholera vibrio is not encapsulated and does not form spores; it is Gram-negative and stains with aniline dyes.

Cultivation characteristics: the cholera vibrio develops within large temperature limits (5—40°C), in the conventional media, especially alkaline, that do not favour the development of other germs. On

TABLE 8.1.

Main characteristics of the cholera vibrio

Morphologic characters	
– fresh sample	In the pathologic product (stool) the cholera vibrio appears in the form of short (1.5 – 3 μ), bent, motile rods, single or united end-to-end in spirals; accentuated pleomorphism under certain conditions. Non-sporulate, non-capsulated, the vibrio moves swiftly across the microscopic field, alone or in clusters like a shoal of fish. In cultures, in terms of the age of the latter, the vibrio presents typical forms up to long or granular, bacillus-like forms.
– stained sample	Stains slightly with all aniline dyes, is Gram-negative and takes on various shapes (straight or bent rods, spiral-like forms).
Cultivation characteristics	
	Aerobic, occasionally anaerobic (in the presence of glucose). Readily grows on conventional isolation media, optimal pH 8.6 – 9.2, at temperatures ranging between 5° and 40° (optimum 37°).
– fluid media	Develops rapidly, especially in peptone water; at two hours dead waves arise on stirring and at 10 hours a film forms and opacity increases. In broth the film does not form and the fluid becomes turbid after 5 – 6 hours.
– solid media	At 8 – 10 hours the colonies are fine, transparent, bluish, measuring 1 mm in diameter; later on they extend over a wider area. S, R (patients treated with antibiotics, carriers), and folded colonies, due to the secretion of a mucous substance (but which readily return to S forms) develop.
Metabolic characters	
	Milk is coagulated only by freshly isolated vibrios. Litmus milk is acidified. Gelatin and coagulated serum are liquefied. Gelatin has been introduced in several isolation media for the characteristic liquefying aspect on slides: the colonies appear to be surrounded by a zone of liquefaction, similar to a shiny glass bead when examined with the magnifying glass.

The vibrio attacks some glucides – glucose, fructose, laevulose, saccharose, maltose, glycerine, mannite, mannose – without the formation of gases; it has no action on xylose, dulcite, arabinose and inulin. Salicin is rarely attacked, lactose more frequently but late. According to Heiberg's scheme it may be included in group I: saccharose + mannose + arabinose.

It develops on media with citrate in less than 10 days.

It reduces methylene blue, does not reduce neutral red, transforms nitrates into nitrites and produces indol, hence the known *cholera rot* reaction on adding sulphuric acid to the culture prepared with peptone water, reaction that is not, however, specific

Produces H_2S; the methyl red test is negative and the Voges-Proskauer test served to differentiate the classical from the El Tor vibrio. The cholera vibrio has a rich enzymatic equipment: mucinase, cholagenase, lipase, elastinase, catalase, oxidase, cytochromoxidase

The classical vibrio haemagglutinates some red blood cells (sheep, goat) but not chicken red blood cells as the El Tor strain does; haemolysis differentiates the two types, the classical strain acting upon human and rabbit red blood cells but is inactive upon goat and sheep red blood cells. Apart from the minor antigens, the cholera vibrio possesses the major A, B, C antigens.

the surface of certain media the cholera vibrio tends to form films; on solid media it forms small colonies, characterised by their transparency in comparison to other enteric bacteria. From patients *S* forms have been isolated, from carriers *F* forms, which raises diagnostic problems, as well as the *R* forms that can be isolated only in special media.

Folded colonies, dwarf colonies, etc., have also been described.

Biochemical features (Table 8.2) are particularly important for identifying the cholera vibrio and its types.

According to Heiberg's (1934) scheme based upon the behaviour of the vibrios to sugars, they ferment mannose and saccharose but not

TABLE 8. 2.
Cholera vibrio differentiation tests

Tests	Classical cholera vibrio	El Tor cholera vibrio	NAG cholera vibrio
Voges-Proskauer	–	+ (recently isolated strains)	Variable
Haemolysis of sheep or goat red blood cells	–	+ (most frequent)	Variable
Agglutinates chicken red blood cells	– (most frequent)	+	
Gipsen (agglutinability of the suspension heated 3 h at 56°)	–	+	–
Tanamal (agglutination with O serum diluted 1 : 200 in the presence of 0.3% sodium carbonate	–	+	–
Precipitation in the presence of mercury chloride	–	+	
Cultures in media with trypsin (broth with 9 parts 1% trypsin)	– (no turbidity)	+ (turbity, freshly isolated strains)	–
Sensitivity to Mukerjee group IV bacteriophage	+	–	Variable
Sensitivity to Polymixin B	+	–	–
Growth in a medium with hexamine	–	+	

arabinose, which lists the vibrio cholerae in group I. However, there are exceptions which list them in group III. It was found that 10% of 258 classical cholerae vibrios ferment arabinose in paraffin sealed tubes although in open tubes fermentation is transitory and delayed.

It should be recalled that some vibrio strains break down lactose; as the process is slow and delayed there is no difficulty in differentiating them from *Escherichia* within the first 24—48 hours.

In the case of El Tor vibrios the Voges-Proskauer test is positive; however, some older cultures no longer produce acetylemethylcarbinol.

The action of the vibrios on the red blood cells has been studied in detail and represents an important criterion of differentiation. The behaviour of the vibrios differs with the type of red blood cells used in the laboratory tests; with goat red blood cells the classical vibrios are considered haemodigestive and El Tor haemolytic. Today however, more and more El Tor strains are proving non-haemolytic on isolation, yet the strains selected on the basis of their sensitivity to bacteriophages acquire haemolytic properties when cultivated in media with heart extract and glycerine.

Haemagglutination of chick red blood cells appears to be characteristic of El Tor vibrio, but the classical vibrio also has a haemolytic action on other types of red blood cells.

Antigenic structure. Being a mobile germ the cholera vibrio has 10 flagellate, protein, unspecific *H*-antigens and somatic *O*-antigens; three of these are major antigens (*A* principal, *B* and *C* accessory), the others are considered minor antigens (*D*—*M*).

According to Gardner and Venkatraman (1935), the cholera vibrios form part of the *OI* group and always contain the antigenic *A* factor; according to the presence of *B* and *C* antigens the cholera vibrios are included in one of the three serologic types identified by Japanese authors (Table 8.3):

TABLE 8. 3.

Ogawa type	AB	
Inaba type	AC	
Hikojima type	ABC	
—	A	not recognised by any investigator

According to the presence of the other minor antigens, several different antigenic formulae of importance for the epidemiological investigations may exist within the serological types.

Recent observations have shown that frequent antigenic changes occur within the same cholera focus and in the same patient from whom different serological types may be isolated successively or alternately.

The antigenic structure is one of the means of identifying the cholera vibrios by agglutination with sera of the *OI* group or with monospecific sera. The vibrios not agglutinated by the specific serum are considered non-agglutinable vibrios.

The non-agglutinable (NAG) or non-cholera vibrios (NCV) were isolated from clinical cholera-like cases but their detection did not necessitate the compulsory control measures applicable when the classical or El Tor vibrio is isolated.

Certain antigenic relationships exist between the antigenic *O*-factors and the antigens of certain bacteria of the *Enterobacteriaceae* family, *Brucella* genus, etc. so that strictly specific diagnostic sera have to be prepared.

Two protein antigens have been extracted from *V. cholerae* that stimulate the formation of vibriocide and, respectively, antitoxic antibodies; these antigens might represent a basis for the preparation of a new immunising antigen.

Numerous detailed studies have been carried out on cholera vibrio bacteriophages, isolated from the faeces of patients, from latrines, water and from lysogenic strains.

The cholera bacteriophage was considered an important factor of variability and Doorenbos (1932) described several variants linked to the different characters of the *V. cholerae* (haemolysis, agglutinability, etc.). D'Hérelle (see d'Hérelle *et al.*, 1930) considered that the transformations induced under the influence of bacteriophages are irreversible, whereas Doorenbos sustained that the atypical forms may return to their initial form and thus play an important epidemiological role.

Gallut and Nicolle (1963) described six lysotypes and found they depended upon their geographical origin; of the 8 phage preparations, No. 1 and 2 are active only on the El Tor vibrio.

In India, Mukerjee (1963 a, b and c; 1970) and his co-workers devised a simplified lysotyping scheme (Table 8.4) that can be used for the diagnosis. The bacteriophages isolated by Mukerjee were divided into four groups in which only 1% of the 4066 classical vibrio strains tested could not be classed.

Between the lysotype and the serologic subgroup or pathogenicity of the vibrio no relationship could be established. However, in Calcutta

TABLE 8. 4.

Lysotyping scheme of *V. cholerae*

Classical cholera phage type	Sensitivity to group phage			
	I	II	III	IV
1	+	+	+	+
2	−	+	+	+
3	+	−	+	+
4	−	−	+	+
5	+	+	−	+

(After Mukerjee, 1970)

cholera vibrios type 1, prevalent during the 1955—1966 period, were replaced by type 3 and spread from there to the neighbouring areas.

Part of the classical vibrio strains are lysogenic but no classification could be drawn up on the basis of this property.

Mukerjee likewise showed that El Tor vibrios are resistant to the action of group IV bacteriophages and thus was able to develop a method for differentiating it from the classical vibrio (Monsur *et al.,* 1965).

Mukerjee's method appears superior to the other methods since it also identifies non-haemolytic El Tor vibrios and presents no variations in time (with other tests the variations may range up to 30%). Mukerjee established the scheme shown in Table 8.5.

Monsur *et al.* (1970) demonstrated, however, that this resistance is not absolute but depends upon the concentration of the bacteriophage, which imposes the use of critical dilutions, with quantitation of the resistance (or sensitivity) to the bacteriophage.

Rizvi and Benenson (1966) obtained, in the laboratory, classical vibrio strains resistant to bacteriophage group IV but maintaining all the other characteristics; if the phenomenon also occurs in nature

TABLE 8. 5.

Lysotyping scheme of *Vibrio El Tor*

El Tor phage type	Sensitivity to group phage				
	I	II	III	IV	V
1	+	+	+	+	+
2	+	+	+	−	+
3	+	+	−	+	+
4	+	+	−	−	+
5	−	+	−	−	+

(After Mukerjee, 1970)

then the question arises of the validity of the criterion of classification as well as that of the efficiency of the bacteriophage treatment, which might only change the lysotype.

Basu and Mukerjee (1970) isolated bacteriophage ∅ H74/64, designated by them as bacteriophage group V, which produces specific lysis of the pathogenic El Tor strains. The classical cholera vibrios, the non-pathogenic El Tor vibrios and the NAG vibrios are not sensitive to this bacteriophage nor to 100 RTD (only exceptionally).

Of practical interest is likewise the finding of Takeya *et al.* (1967 a), i.e. that the pathogenic El Tor vibrios can release a bacteriophage (kappa) which can be tested on a classical vibrio indicator strain (H 218 Sm^r); the El Tor vibrios isolated from water or from very attenuated cholera cases (Ugol strains) do not appear to be lysogenic.

Takeya, Shimodori and Gomez (1967 b) established a correlation between the cultures for the isolation of El Tor vibrios and the detection of kappa bacteriophages in contacts and patients; in the latter, during the onset the vibrios are in far greater numbers (10^6-10^4 germs ml^{-1} as compared to 10^2-10^4 bacteriophage particles). The antibiotic treatment leaves only the bacteriophage as an aetiologic diagnostic element.

Detection of the kappa bacteriophage may represent a diagnostic method for carriers or patients when the vibrios can no longer be found or are covered by an associate flora or NAG vibrios.

Cholera bacteriophages were used for therapeutical purposes by d'Hérelle but their efficiency was questioned (see d'Hérelle *et al.*, 1930).

Dutta and Panse (1963) demonstrated experimentally in the newborn rabbit that the bacteriophage prevents the characteristic manifestations of the disease if administered before the vibrios and lowers the mortality rate if used within the first 8 hours of infection.

Pasricha *et al.* (1936) likewise observed a lower case fatality rate among the patients to whose treatment he had associated bacteriophage.

In Pakistan, Sayamov (1963) found an improvement in the condition of the patients receiving bacteriophage and a lower number of admissions in the localities in which it was administered preventively. Similar good results were recorded in Afghanistan where 270 000 people were treated with bacteriophage.

The method did not become widespread, however, because of the difficulties of preparing an efficient product and of repeated mass administration.

Resistance of the vibrios to the action of physical and chemical factors is particularly important from the epidemiological point of view.

At 56°C the vibrios are destroyed in 30 minutes and, by boiling, in a few seconds. The cholera vibrio supports low temperatures very well: in the refrigerator (+1 to +4°C) the cultures remain viable for 4—6 weeks, and (according to Pollitzer, 1959) some strains may be kept for four months on ice. At very low temperatures survival of the vibrios is greater at – 70°C than at –20°C, in terms of the rate of congealing, the initial concentration of the germs and the presence of protective substances (sugars, fat-free milk, meat extract).

Dryness and sunlight kill the vibrios within a few hours and ultraviolet light within 10 to 45 minutes.

Survival of the vibrios in the environment depends upon the temperature, pH, content of mineral salts or organic material, solar irradiation, etc. as well as on the respective strain.

At pH = 5 a very small number of vibrios survive, whereas in alkaline pH (8.6—9.2) they resist very well. Due to their acidity, the gastric juice and, to a lesser extent, the saliva represent the first barriers in the defence of the organism against cholera.

As regards the mineral salts, a concentration of 250—700 mosm at a pH of 6—8 proves ideal for the survival and multiplication of the vibrios, especially as this osmotic pressure is to be found in most foodstuffs, corresponding to a concentration of 0.5—2% NaCl and a somewhat lower proportion of potassium salts.

In *surface waters* the cholera vibrio survives from a few days up to three weeks. Pandit *et al.* (1967) demonstrated that the duration of survival varies in terms of the water source (in wells rather than in the town water mains of New Delhi) and temperature (only 1–4 days at 37°C as compared to 18–51 days at 21°C). Following a single contamination and daily use of the well the survival was of 10–12 days.

Pesigan *et al.* (1967 a) also observed the longer duration of survival at low temperatures. From the data of Felsenfeld (1965) and Neogy and Mukherji (1970) it results that the classical strain resists less than the El Tor strain and that survival in water depends upon the pH (Table 8.6).

TABLE 8. 6.

pH of water	Classical vibrio	El Tor vibrio
6.4–6.6	1–2 days	8 days
7.6–8.8	7.5±1.9 days	19.3±5.1 days

In the water supply system of towns survival is of, at most, one hour due to chlorination.

In their experiments, Pandit and co-workers did not detect any influence of the associated flora, although in the laboratory combined cultivation demonstrated that pyocyaneus may cover the growth of El Tor vibrio. Gallut and Quinon (1970) showed that the classical vibrio is overtaken by the development of El Tor vibrio, and both strains by the growth of NAG vibrios, a phenomenon which they consider to be involved in the present epidemiological trend of cholera and which accounts for the end of an epidemic with the appearance of NAG vibrios in the choleraform syndrome of the latter period.

This microbial comptition must be borne in mind when attempting to isolate the vibrios and demands the use of selective media.

Survival of the vibrios in the *faeces and cadavers* is, within the limits shown, influenced by temperature, humidity and microbial competition (Forbes *et al.*, 1967; Kenneth *et al.*, 1970). No comparative studies have been carried out on survival of the classical and El Tor strain in the faeces but the latter is more frequently isolated from latrines and in greater amounts even where there is a concomitant outbreak of both El Tor and classical cholera (Bart *et al.*, 1970). To this situation may also contribute the asymptomatic forms of cholera, more frequent in the case of the El Tor vibrio, and which spread the germ widely.

Survival of the vibrios in contaminated food — a problem with broad epidemiologic implications — has been studied in numerous investigations. The data of Pollitzer (1959), Felsenfeld (1965), Prescott and Bhattacharjee (1969) have shown the following:

(i) the El Tor vibrios survive 2.5 times longer than the classical vibrios;

(ii) foods with a high sugar content (more than 5%) and an increased osmolarity (5% sodium chloride) permit only a short survival;

(iii) in foods rich in organic matters and water, survival is longer;

(iv) storage of contaminated food in the cold prolongs the survival time of the vibrios; in the refrigerator fish remains contaminated for 9 days, and at −20°C the vibrios may last for up to a month in meat and fish;

(v) cholera vibrios rapidly disappear from vegetables and fruit with an acid reaction (a pH of less than 5 is very harmful).

Further data show that cholera vibrios survive:

(i) in milk, milk products, sweetmeats, flour pastes up to 4 weeks;

(ii) in intact raw vegetables and fruit kept in the refrigerator up to two weeks but at room temperature less than a week. Prescott and Bhattacharjee (1969) showed that at 20—25°C the survival of vibrios is:

1 hour in lemons;
1 day in oranges;
2 days in bananas;
1 hour in figs;
1 day in tomatoes;
1 day in grapes and dates;
3 days in onion and peas;
8 days in potatoes;
3 days in wheat and barley;
1 — 5 days in spices;
1 day in coca-cola;

Longer survival was observed in slices of melon, carrots and cooked eggplant.

It is worthy of note that the pulp of intact fruit and vegetables was not contaminated, invalidating the theory of the penetration of vibrios through the roots. Therefore, outer decontamination is sufficient; vegetables kept for one minute in boiling water kills all the vibrios, but other methods with various products also exist.

A further idea that results from these findings is the lack of infectivity of imported fruit (lemons, oranges, bananas, figs, dates, spices, etc.) when the transport time exceeds the limit of survival of cholera vibrio; the risk is non-existent and there is no point in restricting trade with endemic countries.

On other objects, survival of the vibrios is less than one week: on paper, plastic, metal, coins, vibrios may persist for 3 days, and in kitchen utensils and cutlery about 1 or 2 days. On tissues they last a few days or up to, at most, one week.

The list of survival of cholera vibrios (Table 8.7) was published by Barua in 1970.

TABLE 8. 7.

Product	Survival in days of cholera vibrios at temperatures of:	
	30–32°C	5—10°C
Food		
Fresh vegetables	1–7	7–10
Fresh fruit (including oranges)	1–3	3–5
Dry fruit (dates, figs, hazelnuts)	1–3	–
Peas, lentils, rice, wheat	1–3	3–5
Dairy products	7–14	over 14
Fish (smoked, salted, dried)		7–14
oysters	2–5	(more than 21 if congealed)
Cooked food	2–5	3–5
Sweetmeats	1–2	
Drinks (beer, soda water, coca-cola)	1	1
Spices	1–5	
Varia		
Ground coffee, tea leaves, yoghourt	at most 1 day	
Rice in water	1 hour	
Water in reservoirs	7–13	more than 18
Sea water	10 13	60
Metal objects, porcelain	1–2	
Cloth, leather, rubber	3–7	

(after Barua, 1970 a)

The action of various disinfectants. The cholera vibrio is extremely sensitive to a series of disinfectants and is destroyed more rapidly than other intestinal bacteria.

Table 8.8 gives the sensitivity of cholera vibrios to various disinfectants:

TABLE 8. 8.

Disinfectant	Concentration	Inactivation time in minutes		
		El Tor vibrio	Classical vibrio	Typhoid bacillus
Phenol	1/100	2.5	2.5	15
Cresol	1/300	15	15	
	1/380	2.5	2.5	15
	1/400	5	10	
Benzetonium	1/5000	5	5	10
chlorate	1/10 000	15	15	
Benzalkonium	1/5000	5	5	5
chlorate	1/10 000	15	10	15

These data show that the cholera vibrios are at least twice as sensitive as the typhoid bacillus. The El Tor vibrio is considered to have a greater resistance than the classical vibrio.

Further data show that:

(i) mercury bichloride in a 1/100 000 concentration destroys the cholera vibrios in less than 5 minutes;

(ii) in sodium hypochloride and lime chloride vibrios resist very little; limewash (0.5 g powder / litre with 38% chlorine) destroys the vibrios in 10 minutes;

(iii) ethylene oxide depends upon the concentration and temperature at which it acts: in a 10% concentration it destroys the cholera vibrios in 6 hours at 30° and in less than 1 hour at 40°;

(iv) potassium permanganate in a 0.2 g/litre concentration disinfects in 15 minutes;

(v) 10% alcohol readily destroys the vibrios.

Sensitivity of the cholera vibrios to chemotherapeutical drugs and antibiotics

Many investigations have been conducted in this connection. The results obtained *in vitro* were compared to those obtained in the treatment of patients and preventive administration to contacts (Gangarosa *et al.*, 1966; Kobari *et al.*, 1967 a and b; Kuwahara *et al.*, 1967; Pesigan *et al.*, 1967 b; Uylangco and Vasco, 1967 a).

It was demonstrated *in vitro* that tetracyclin, chloramphenicol and erythromycin have a high bacteriostatic activity, then streptomycin, kanamycin and novobiocin which are moderately active; spiramycin, oleandomycin and benzylpenicillin are only slightly bacteriostatic. Cephaloridin, cephalotin, ampicillin and sulphaphen-azol gave variable results.

The following are some of the published results:

Tetracyclin inhibits 60% of the El Tor vibrios at a concentration greater than 0.78 μg ml^{-1} and 81% at 3.12 μg ml^{-1}; the classical NAG strains do not resist a concentration higher than 0.78 μg ml^{-1}.

Chloramphenicol is very active in concentrations of 1—2 μg ml^{-1}; most of the El Tor strains are inhibited at concentrations higher than 3.12 μg ml^{-1}, whereas no classical and NAG vibrio strains resist at these values.

Erythromycin as a rule active in 6.25 μg ml^{-1} concentrations suppresses 90% of the strains at more than 12.5 μg ml^{-1}.

Streptomycin, *kanamycin*, *neomycin* inhibit the vibrios in a wide range of concentrations (6.25—50 μg ml^{-1}).

Novobiocin inhibits 90% of the vibrio strains in concentrations higher than 6.25 μg ml^{-1}.

Polymixin exercises an effect on El Tor and NAG vibrios only in a dose higher than 250 U ml^{-1}, similar to colistin sulphate (active in concentrations of at least 1 000 U ml^{-1}) and colistin methane sulphate (5 000 U ml^{-1}). On the other hand, the classical vibrio is sensitive to 2.5—100 U ml^{-1} polymixin, hence sensitivity to this product has been used as a common test for identifying the vibrio type.

From the findings published to date it seems that:

(i) The classical vibrios are more sensitive to antibiotics and the NAG strains are at least as resistant as the El Tor ones.

(ii) In the same focus no differences were noted between the sensitivity of the strains isolated from the treated and untreated patients, except for cases treated with kanamycin when the strains present a cross resistance to streptomycin.

(iii) There was no initial difference in sensitivity according to the origin of the strains; but this now becomes evident with chloramphenicol and tetracyclin. Prescott *et al.* (1968) explains the greater resistance of the strains from Calcutta than of those from Thailand and Philippines by the widespread use of antibiotics in India.

(iv) The more resistant strains appear to be isolated from latrines.

(v) Resistance to antibiotics decreases by repeated passages in culture media parallel to diminution of the enzymatic activity.

Attention has been drawn to the increasingly frequent isolation of vibrio strains resistant to the currently used antibiotics chloramphenicol, tetracyclin and streptomycin which will raise difficult problems of curative or preventive treatment with antibiotics.

The cholera vibrio frequently acquires rapid resistance to streptomycin. Kobari *et al.* (1970) showed that the strains isolated in the Philippines are generally sensitive to 12.5 — 25 $\mu g\ ml^{-1}$, but that numerous strains were not inhibited even by 100 $\mu g\ ml^{-1}$.

To chloramphenicol concentrations higher than 25 $\mu g\ ml^{-1}$, 6% of the strains were resistant and 2% of the strains resisted to tetracyclin administered in the current daily doses.

A change in the sensitivity to certain antibiotics appears to be correlated to the transfer of a resistance factor. This may account for the increased resistance to chloramphenicol observed in Manilla before the use of this antibiotic.

Transfer of the *R* factor from some enterobacteriaceae to the cholera vibrio and from vibrio to vibrio or from vibrio to other enterobacteriaceae has been proved experimentally.

Prescott *et al.* (1968) found evidence of the presence of the *R* factor in four of 225 cholera strains isolated from patients.

In conclusion, from the published data it seems that of the widely used antibiotics, tetracyclin, chloramphenicol and erythromycin continue to be active against cholera vibrios, and the proportion of resistant strains in the countries with endemic cholera is still very low.

In the intact countries, an imported focus may be triggered by resistant strains; this points to the necessity for determining the sensitivity of the incriminated vibrio at least in the initial period of the epidemic.

The cholera vibrios have proved sensitive to some chemotherapeutical drugs.

Sulphonamides are used in many countries but do not have a spectacular action on vibrios. The minimal inhibitory concentration of Sulphafurazol is about 10 $\mu g\ ml^{-1}$ against approximately 50% of the El Tor strains. The classical vibrio appears to be resistant.

Furalazin and *furazolidin* — nitrofuran derivatives — have proved very active against the cholera vibrios; their minimal inhibitory concentration is of 0.047 $\mu g\ ml^{-1}$ and respectively 0.39—1.56 $\mu g\ ml^{-1}$. However, a diminution in the sensitivity of the vibrios to Furazolidin has already been observed. During the 1962—1967 period the mean sensitivity changed from 1 $\mu g\ ml^{-1}$ to 8 $\mu g\ ml^{-1}$.

Observations on the sensitivity of cholera vibrios to antibiotics and chemotherapeutics, determined *in vitro* were confirmed in the clinic.

The efficiency of the drugs was assessed by their influence on the duration of diarrhoea, the amount of fluid lost, saline substitution requirements, the duration of vibrio excretion and bacteriological relapses.

Clinical practice has shown the superiority of tetracyclin and chloramphenicol; although very active erythromycin is too expensive.

Several therapeutical schemes have been tested; the most widespread is the administration of 500 mg every 6 hours, for 2 days.

The duration of diarrhoea is reduced in general to 2 days as compared to the 4 days of cases not treated with antibiotics; according to Kobari *et al.* (1967 a and b) the frequency and volume of the stools only begins to diminish 12 hours after beginning the treatment.

In the treated adult the volume of the stools eliminated is on average 8 as against 21.5 litres in the untreated cases. Both in adults and children the substitution fluid may be reduced by at least 50%.

Elimination of the vibrios is about 2–3 days instead of 5–6 as in the untreated cases, especially after the administration of tetracyclin. A fall in the number of excreted vibrios is very sharp from the first hour of administration while diarrhoea continues probably because the function of the intestine recovers more slowly.

However, there is a reverse to this prompt response to antibiotic treatment: the proportion of bacteriological relapses averages 20% as against 8–10% in patients not receiving tetracyclin or chloramphenicol.

The administration of sulphaguanidin has given satisfactory results (Wallace *et al.,* 1968), but the effect is not considered to be the same (Lindenbaum *et al.,* 1967 a and b). Gangarosa *et al.* (1967) recommended the use of long term sulphonamides.

In 1970, Gharagozloo *et al.* used trimethoprim-sulphamettoxazol with good results.

Furazolidin was administered in 250 mg doses every 6 hours for 3 days; in 4–12 year old children the dose was reduced by half and in children under the age of to 5 mg kg^{-1} body weight per day. The results were similar to those obtained with tetracyclin and chloramphenicol. This product is recommended especially in cases with antibiotic-resistant vibrios (Chaudhuri *et al.,* 1968).

Finkelstein and Raungthum (1968) noted the favourable effect of enterovioform (Clioquinol) administered experimentally to the rabbit in a single dose.

The Indian workers have shown that berberina (an alkaloid extract of *Berberis aristata*) inhibits the cholera vibrio in a 1/5 000 dilution and modifies experimental cholera; its administration in the hospital of Calcutta sometimes proved more efficient than chloramphenicol.

In conclusion, the treatment with antibiotics appears to be just as efficient when administered *per os* as by the intravenous route; actually, the latter is recommended for starting the treatment before vomiting has ceased.

Non-absorbable antibiotics do not seem to be more efficient because they are eliminated so rapidly that contact with the intestine lasts for only a short time.

Chloramphenicol has not proved as efficient in children as in adults, probably because it is administered in a palmitate syrup, the inert form of the antibiotic which cannot be hydrolysed because of the rapid intestinal transit.

Initial dehydration and denutrition influence the duration of diarrhoea, irrespective of the antibiotic therapy, which does not improve the circulatory collapse.

The use of small antibiotic doses does not give favourable results eliminating vibrios from the intestine.

In the control of a cholera focus in intact countries it is recommended to prolong the antimicrobial treatment up to 7 days in order to be certain of the total elimination of vibrios.

The *variability* of the cholera vibrio may raise problems concerning the bacteriological, epidemiological diagnosis and therapeutical possibilities.

Marked changes have occurred in the morphologic, cultural, metabolic characters and pathogenicity of the cholera vibrios.

Among the chief factors of variability is the bacteriophage, chlorine used for the treatment of water (accounting for the different properties of the strains in the endemic foci) and the action of antibiotics.

Transformation in the antigenic structure of the vibrio has frequently been reported, isolating various serotypes from the same patient, successively or even concomitantly.

Gangarosa *et al.* (1967) isolated all three serotypes (Ogawa, Inaba, Hikojima), from the same person at first, successively, and then in convalescence simultaneously, the other characters being identical, while in Iran in 1965 cholera of the Ogawa type evolved. Similar observations were made in Taiwan in 1962. From a laboratory agent, accidentally spattered with an Inaba vibrio culture, the Ogawa type was subsequently isolated. Gangarosa considered that in relapses a new serologic type may be isolated.

Changes in the serotype have been obtained in the laboratory and are attributed to the appearance of antibodies in the organism in the course of infection.

The same mechanism also stands at the basis of the morphologic changes expressed by the appearance of *R* forms 15 to 29 days after the onset of the disease. These forms seem to be better adapted and more resistant. The *R* forms are frequently encountered in long-standing carriers. The epidemiologic consequences and diagnostic problems are obvious.

In this category of changes are also those of the *L* forms, which can also be obtained experimentally. The fact that the *L* forms only grow on special media and after a longer interval (48 hours) may explain the negative cultures of the typical cases in cholera foci.

The spread of El Tor and NAG vibrios is not yet fully understood. It has been assumed that the intervention of certain environmental factors, including the presence of chlorine in the water, have imprinted new biochemical characteristics on the cholera vibrio; a similar role would be that of the circulation of the vibrio in a partly immunised population. However, it has not been possible to demonstrate experimentally transformation of the classical vibrio into the El Tor vibrio or the converse, although both biotypes have been frequently isolated from the same patient.

El Shawi and Thewaini (1969) isolated NAG vibrios that can be classed in Heiberg's group I since in the laboratory they took on the fermentation properties characteristic of this group. Their role has not been made clear; they have also been isolated from diarrhoeic syndromes and, following passages in suckling rabbits, have become agglutinable. This instability of NAG vibrios has drawn attention to their untoward epidemiological potential.

Differentiation of the classical from the El Tor vibrio cannot be done on the basis of morphologic, cultivation or antigenic criteria; nor does their pathogenicity differ. At present they are differentiated on the basis of their metabolic characteristics and bacteriophage-bacteria relationship.

Although numerous methods of differentiation have been devised the most important are haemolysis, sensitivity to hexamine and polymixin B, and their behaviour with regard to bacteriophages of groups IV and V on Mukerjee's scheme.

Most investigators in India consider their action on goat red blood cells as a basic criterion; the haemolytic potential absent in the classical strain is present in El Tor.

Mukerjee (1970) considered the El Tor strain resistant to group IV bacteriophage and subsequently characterised them by their sensitivity to group V bacteriophages; the classical strain has an inverse behaviour but as they may sometimes be resistant to group IV bacteriophage this criterion becomes questionable.

Neither is the haemolytic potential of El Tor vibrios quite clear. Numerous strains exhibiting the properties of El Tor in other tests, of which resistance to group IV bacteriophages has proved to be non-haemolytic. In India many investigators prefer to consider cholera vibrios as atypical.

Korobkova and Zukin consider haemolysis rather as a quantitative than a qualitative differentiation test. Following dissociation of young strains that do not produce haemolysis, both non-haemolytic and haemolytic colonies could be obtained; infact, it is considered that the ratio of the former to the latter must be 1 : 5 for the isolated strain to manifest haemolytic properties.

It is certain that an increasing number of isolated El Tor strains do not manifest haemolytic properties. For instance, in 1961, the El Tor vibrio that invaded the Philippines was haemolytic but this property subsequently became weaker and then non-haemolytic strains gradually appeared.

It has been reported that 1% glycerine in broth with brain and heart extract reveals the haemolytic properties of El Tor vibrios considered as non-haemolytic.

One of the explanations of this change in the haemolytic properties is the exposure of the vibrios to the action of antibiotics that inhibit protein synthesis, a phenomenon not yet encountered with the antibiotics that affect the cell membrane or wall.

Moreover, neither does the classical vibrio lack haemolytic properties completely since it acts upon human and rabbit red blood cells.

A similar aspect is encountered in the case of haemagglutination, a property that certain classical strains acquire after being carried through several passages on culture media or rabbit intestine.

Nor is the Voges-Proskauer reaction an absolute criterion as some El Tor strains do not produce acetylmethylcarbinol.

The hexamine test appears to be one of the most constant and certain: the El Tor vibrio may develop and attack glucose in the presence of a concentration of 0.125% hexamine. Similarly constant is the sensitivity test to polymixin B, but account must be taken of the fact that at certain concentrations even the classical vibrio is resistant. The use of colistin has been advocated: discs with 50—100 units polymixin B, 150 units colistin sulphate and 1 000 units colistin methanesulphate.

At any rate, testing to these antibiotics should be done under strict standardised conditions. The Mueller-Hinton medium and standard discs with antibiotics (polymixin B, colistin sulphate) have been specially recommended.

On Monsur and TCBS media up to 85—100% of the El Tor vibrios might be considered as classical if the zones of inhibition around the antibiotic disks are taken into consideration.

The pH of the medium must be controlled: between 7.8 and 8 a false zone of inhibition appears around the El Tor colonies, It is also recommended that a strain of known sensitivity and one of known resistance should be tested at the same time.

8. 1. 2. Pathogenicity of the cholera vibrio

Cholera is a natural disease of man; the cholera vibrio has not been found in any other animal species.

The need to investigate the pathologic physiology of cholera and therapeutical possibilities has resulted in the choice of experimental models as close as possible to natural conditions. At present the newborn rabbit, the isolated intestinal loop of the adult rabbit and dog are used almost exclusively.

In cholera, the vibrio remains in the intestine where it multiplies without spreading to the rest of the body. For some time it has been assumed that the liberation of toxic products at this level causes the manifestations of cholera (Oza and Dutta, 1970).

Indeed, numerous studies have revealed several toxic factors that produce, in the experimental models mentioned, alterations identical to those obtained following inoculation of the cultures as such and to those occurring in human cholera: passage of fluid into the intestinal lumen (young rabbit, the isolated loop of an adult rabbit), increase in vascular permeability (by intradermal inoculation), inhibition of sodium transport, intoxication phenomena affects of the endotoxic type.

The problem of these toxic factors has not yet been fully elucidated, many research workers still denying that the cholera vibrio produces an exotoxin. Hence, Burrows (1968) preferred the term *toxic activity*. This activity was observed both in cholera vibrio cultures (in their supernate or by lysis of the vibrios) and in the patients' stools.

According to Burrows the toxic activity of cholera vibrios can be separated into three groups according to thermostability, dialysis, diffusion from the cell, and to a lesser extent by its location in the microbial cell. He proposed to call these groups *toxin types*, eventually with subtypes.

Type 1 toxins correspond to the conventional endotoxins, situated both within the cell wall and intracellularly; they are thermostable and do not dialyse through cellophane. The endotoxin is released by autolysis, kills the chick embryo (0.01 μg ether extract by intravenous route) and produces a haemorrhagic reaction on the chorioallantoic membrane of the embryonate egg.

Type 2 toxins are to be found intracellularly but not in the cell wall. They appear in the supernate of the cultures, are thermolabile (56°C, for 15 minutes) and non-dialysable. Type 2 toxins have two subtypes: choleragenic toxins (produce cholera-like illness in animals) and toxins demonstrable by other methods (intradermal inoculation or in cell cultures), i.e. toxins 2a and 2b.

Choleragen toxins do not appear with all strains and under any conditions; special media are recommended for their preparation.

The choleragen material produces phenomena similar to cholera when administered by intubation to suckling rabbits, in the isolated intestinal loop of the rabbit, and in the dog.

During the last few years numerous studies have been conducted for purification and concentration of these enterotoxic factors, especially on Sephadex columns.

Toxins 2a correspond to the *choleragen* described by Finkelstein (1969). It has a molecular weight of 61 000 daltons and is assumed to be of protein nature, non-dialysable, thermolabile.

Not all the strains produce choleragen *in vitro* but *in vivo* their number increases; the fact that the choleragen is not always identified may be accounted for by the fact that some strains produce very small amounts or "bonded" forms.

Although it has been assumed that it acts by increasing the permeability of the intestinal blood vessels it has recently been postulated that the choleragen stimulates secretion of the intestinal cells.

By inoculation of animals the purified choleragen stimulates the production of antitoxic antibodies but at the same time vibriocide and agglutinating antibodies appear, probably due to the impurity of the material.

The properties of the preparations (toxicity, potential to combine with antitoxin, immunogenic potential) can be dissociated by lyophilisation, heat, 1% glutamate, formaldehyde.

Other type 2 toxins are demonstrated by:

(i) The results of intradermal inoculation (induration, erythema, increased capillary permeability) expressed in blueing units. The culture supernate contains 10–40 μ ml^{-1} and the faeces filtrate 100 μ ml^{-1}. Craig (1965) calls this activity the permeability factor (PF), an antigenic substance that brings about the production of antitoxin; he considers the PF as identical to the toxin that produces diarrhoea, as also demonstrated by Richardson and Moftle (1970). However, Grady and Chang (1970) obtained an enterotoxin substance lacking the permeability factor.

(ii) Alterations in the cell cultures: the CPE appears within 24–48 hours with the accumulation of lipids in the cells.

Type 3 toxins (thermostable, dialysable) produce inhibition of the sodium pump and are not associated with choleragen toxin.

Haemolysin present in the El Tor vibrio as a lecithinase but does not appear to be identical with the enterotoxin of the vibrios.

From the viewpoint of pathogenicity it is assumed that both types of cholera vibrios may cause all forms of the disease.

Some data appear to show that in El Tor cholera the attenuated, unapparent forms are predominant. The same thing has been demonstrated experimentally: in 1963 Mukerjee showed that the El Tor vibrio is less pathogenic for the suckling rabbit, recently confirmed by Oza and Dutta (1970).

It should be recalled that NAG vibrios were isolated from diarrhoeic syndromes especially towards the end of the epidemics.

●

The role of toxic factors in the pathogenic mechanism of cholera cannot be questioned but their role in anticholera immunity has not been clarified, although many studies have been carried out in this connection.

Once the toxic factors and their role was described the problem arose of their introduction in cholera vaccines or their use in purified preparations.

Nevertheless, the vaccines tested in the Philippines and Pakistan, with satisfactory results concerning the protection conferred, do not produce antitoxins and therefore do not contain these factors.

Toxoides were prepared by heat or formaldehyde, which stimulated the production of antitoxin. Strangely enough the formaldehyde toxoin proved to be a more powerful stimulus than the unmodified toxin. One of these toxoides applied to man induced a serologic response in 60% of the people tested (Feeley and Roberts, 1969).

Practically, the need to use toxoides arises because the toxin inoculated by parenteral route induces no kind of effect.

However, the study of antitoxic immunity is more difficult as the preparations are not sufficiently pure and cause the appearance of vibriocide antibodies.

At present, vibriocide antibodies are considered to indicate a natural or postvaccinal resistance: receptivity to the disease decreases with age, and after vaccination is inversely proportional to the titre of vibriocide antibodies; the presence of the latter does not eliminate diarrhoeic symptoms when the disease has occurred.

The vaccine of the future will probably be both antibacterial and antitoxic.

The study of anticholera antibodies shows:

(i) In convalescents, the response to the choleragen factor and permeability factor is variable and often weak, particularly in the cases treated with tetracyclin (Pierce *et al.*, 1970 a), but remains significant for at least 12—18 months.

(ii) Of interest is the sharp increase in the antitoxic titre in convalescents as compared to the slow and prolonged production of antibodies in the animal inoculated by parenteral route (the local action probably interferes earlier in the former).

(iii) Antibacterial antibodies introduced in the intestine prevent the adsorption of vibrios in the mucosa but do not affect those that are free in the intestinal lumen and which continue to multiply.

(iv) Antitoxic antibodies neutralise the toxin released by vibrios and protect the animal.

(v) Passive antitoxic immunity does not protect rabbits following inoculation of live vibrios, whereas passive antibacterial immunity insures protection.

(vi) The titre of antitoxic and anti-PF antibodies is similar, suggesting their identity, but the level of these titres has no connection with that of vibriocide antibodies.

(vii) Antitoxic antibodies can be titred by neutralisation of the PF. For the study of antitoxic immunity in cholera, this technique is more accessible than neutralisation in laboratory animals.

(viii) Vibriocide antibodies develop following endotoxic stimulation, attesting to resistance of bacterial origin.

Benenson *et al.* (1968 a, b, c, d, e) and Mosley *et al.* (1968 and 1969 a and b) carried out interesting studies on the serologic changes occurring in such situations:

(i) Following vaccination the antibody titre increases considerably especially over the age of 5 years (in endemic cholera regions occult immunity takes place); determination of vibriocide antibodies is more sensitive than that of agglutinins (significant increase 96.5% as against 89.6% in the vaccinates).

(ii) The level of vibriocide antibodies increases at least fourfold in 82% of the family contacts with positive coprocultures.

(iii) Neutralisation of the PF in rabbits shows an increase of the corresponding antibodies in a significant proportion of the confirmed patients (92% show a threefold increase of the titre), comparable to the increase in the agglutinating and vibriocide antibodies. Neutralising activity appears soon (within the first week after the onset) and has the advantage of not being recorded in the vaccinates (at least with present day vaccines).

8. 2. Laboratory tests for the diagnosis of cholera

As a rule the laboratory tests concern the patients and convalescents, the healthy contacts or suspect germ carriers, but environmental factors must also be examined (water, food, soil, sewage and household wastes, etc.).

Whatever the situation the only satisfactory criterion for clinical and especially epidemiological requirements remains the isolation and identification of the cholera vibrio.

In patients, but especially in carriers, serologic investigations reveal the presence of antibodies in the serum, but the results are uncertain and late and hence are not considered sufficiently useful.

In its essence, the diagnosis of cholera must be rapid — if possible from the very first case or cases in an epidemic focus appearing within an intact region. Only thus can the control measures be efficient.

We shall discuss the methods and techniques applied especially in the laboratories of India, a country with a vast experience of cholera (Table 8.9).

8. 2. 1. Products necessary for the bacteriologic diagnosis, collection and transport

The value of laboratory examination depends first of all upon the conditions in which the pathologic samples are collected and handled. Therefore precise indications must be given in this connection.

TABLE 8. 9.

General working scheme for the laboratory diagnosis of cholera

Working stages	Methods for identification of the cholera vibrio	Value of methods
Pathologic product	Microscopic examination of the fresh and stained sample	++
	Specific mobility inhibition	++++
	Immunofluorescence	++
Conservation medium	Microscopic examination of the fresh and stained sample	+
	Specific mobility inhibition	++
	Immunofluorescence	++
Enrichment medium (4, 6,12 h at 37°C)	Microscopic examination of the fresh and stained sample	++
	Specific mobility inhibition	++++
	Immunofluorescence	++++
Selective media (12 h at 37°C)	Morphologic examination of the colonies (oblique light, liquefaction of gelatin, string test)	++
	Microscopic examination of the fresh and stained samples	++++
	Specific slide agglutination	++++
	Immunofluorescence	++++
Differential medium — lactose gelose with bromthymol blue or Istrati-Meitert medium (12 h at 37°C)	Morphologic examination of the colonies (oblique light, liquefaction of gelatin, string text)	++
	Microscopic examination of the fresh and stained samples	++
	Specific slide agglutination	++++
	Immunofluorescence	++++
Gelose 2% (12 h at 37°C)	Morphologic examination of the colonies (oblique light, liquefaction of gelatin, string test)	++
	Microscopic examination of the fresh and stained samples	++
	Specific slide agglutination	++++
	Immunofluorescence	++++
	Biochemical tests	++++
	Sensitivity to bacteriophage	++++
	Sensitivity to polymixin	++++
	Haemolytic activity	++++

The main products are the faeces and vomit, then the clothes and bed linen. From contacts and carrier suspects the faeces are collected, and from convalescents the duodenal content by intubation.

The intestinal content is collected from patients who have died.

For isolation of the vibrio from the environment the faeces are collected from latrines, the sewage and household waste waters, surface or well waters, food samples, fruit, vegetables and flies.

Collection of the stools (Table 8.10)

(1) The fluid stool excreted spontaneously by the patient is the best material for bacteriological examination: it is collected in a container (nightchambers, bedpan, plate, plastic tray, etc.) sterilised by boiling or in a container made from parchment or greased paper.

As a rule the containers in which the stools are collected are disinfected by adequate substances (especially chlorigenic products). As the cholera vibrio is very sensitive to the action of disinfectants, the presence of traces even after the containers have been washed several times is sufficient to give negative results. On the other hand, a sample collected from the container generally used in the cholera annexe may give a falsely positive result.

The containers prepared from parchment, greased paper, plastic or cardboard are disposable and immersed in disinfectant or burnt after use. When other types of containers are used they must be well washed with disinfectants then with running water to eliminate all traces of the disinfectant.

From the patient's stool approximately 1 to 3 ml is taken or 1 g if the stool is less fluid; the samples are collected in special containers or thick glass tubes (Jena test tubes) tightly closed with a rubber stopper.

The containers made of thick glass with a stopper and a scoop are introduced into metal or wooden tubes or, as used in India and recommended by the WHO, glass tubes with a metallic, ebonite or plastic screwed on lid (MacCartney, or similar containers used for drugs).

This type of container has the advantage of a wide opening so that the material can be readily manipulated. They are tightly closed and so cannot leak during transport; having a capacity of 25 ml they may be prepared beforehand in the laboratory with conservation

TABLE 8. 10.

Laboratory diagnosis of cholera: working scheme for examination of the stools (adapte dafter Felsenfeld, 1966)

Stool or swab

↓	↓
Venkatraman and Ramakrishnan medium or alkaline peptone water with tellurite	30% glycerine solution
↓	↓ MacConkey medium, Wilson-Blair medium, Istrati-Meitert medium ↓ Tetrathionate broth (Müller-Kaufmahn)
Alkaline gelose BSA medium MacConkey medium	
↓	↓
TSI medium	TSI medium
↓	↓
Semi-solid medium with mannitol Peptone broth (for indol test) Inclined gelose	Semi-solid medium with mannitol Peptone broth (for indol test) Inclined gelose
↓	
Indol test Voges-Proskauer test Agglutination Haemolysis Haemagglutination Arabinose fermentation Mannose fermentation Saccharose fermentation Lysotyping Other tests (hexamine, sensitivity to polymixin)	Indol test Voges-Proskauer test Agglutination Lactose fermentation Saccharose fermentation Salicin fermentation Decomposition of urea, etc. Utilisation of citrate

or enrichment medium (about 10 ml). This type of flask can be used also for collecting and transporting other products, as found, or in adequate media (conservation or enrichment): vomit, intestinal content (autopsy), linen fragments, the water with which fruit or vegetables are washed, and filtrating membranes through which a large amount of water has passed.

When the laboratory is close to the hospital the stool samples are taken directly, thus facilitating examination (microscopy of the fresh sample and immobility test, as well as immunofluorescence).

If the samples have to be transported farther away a conservation or enrichment medium must be used in order to hasten the sequence of the processing in the laboratory. Choice of the medium depends upon the distance the samples have to be taken; when near by, an enrichment medium is used; when farther away a conservation medium is necessary.

Samples are collected from the patient with a Nelaton type tube or a swab mounted on a metallic or wooden stick. The use of rectal swabs has the advantage of eliminating the vessels for collecting the stools with all the inherent processing and manœuvring, even if the collection itself is somewhat more difficult.

The sterilised rubber tube is kept in a test tube or flask with a screw lid; the swab must have sufficient cotton to absorb 0.1—0.2 ml of fluid. When diarrhoea has ceased the swabs are moistened in peptone water and then pressed out. After collecting, the tube or the swab is "washed" in the fluid of the flask (conservation or enrichment medium). The swab may eventually be left in the flask after the wooden stick has been broken. The swab should not be allowed to become dry as it is then useless (Barua, 1970 a).

Barua and Gomez (1967) recommended another system: to take strips of blotting paper and moisten them in the stools, then pack them in cellophane or plastic envelopes, which are well closed. The blotting paper does not dry and the vibrios survive for 13 to 34 days at room temperature. Filter paper does not give the same results, the vibrios surviving only 3 to 4 days under the same conditions.

This method facilitates collection and transport but care must be taken to pack the envelopes correctly so as to avoid any danger of contamination.

When a great number of examinations are required within a short interval the samples may be pooled, but not more than ten at a time. With increase in the number of samples in the pool the sensitivity of the bacteriological test diminishes (Zafari *et al.*, 1968).

For a bacteriological examination to be effective the *stool sample must be collected before the administration of antibiotics.*

(2) The *faeces of apparently healthy persons* must also be examined (in which case the stools are not fluid). This category includes:

(i) cholera contacts;

(ii) assumed carriers according to certain epidemiological indication or isolation of the cholera vibrio from the latrines used by them;

(iii) individuals coming from countries with cholera;

(iv) former patients.

In such cases the samples are collected either directly by tube or swab or after administration of a saline purge, the chances of isolating the vibrios being greater; for instance magnesium sulphate (15—30 g) in water, (200 ml) the samples being taken from the second stool after the purge. Pierce *et al.* (1970 b) used magnesium sulphate (30—60 g) in water (100 ml); a fluid stool is obtained after 1 to 4 hours after which sampling is done directly with a rectal swab.

Magnesium sulphate is preferred to sodium sulphate because it is cholecystokinetic and empties the gallbladder into the duodenum where there are more chances of finding the vibrios since in carriers the vibrios appear to prefer the bile ducts.

The stool samples are collected with care (preferably in parchment or plastic containers) and introduced into flasks with adequate media, as already mentioned.

For mass investigations pooling of samples from several individuals appears to be indicated; in the event of a positive result the test is repeated individually.

(3) In field epidemiological surveys, especially in rural areas, the faeces are collected from latrines for bacteriological examination. According to the local situation samples are taken from several points or the faeces are mixed with a stick and single samples taken. The samples are introduced into test tubes or flasks, eventually with conservation or enrichment media, i. e. 2 to 3 g well mixed per 50 ml medium.

Collecting the vomit

Vomit is collected in the same way as the stools; adequate amounts are introduced in the containers used for collecting.

Collecting the duodenal fluid

As in carriers, the vibrios seem to prefer the bile ducts, and duodenal intubation is performed.

The duodenal fluid is aspired through the tube after the administration of magnesium sulphate (30 g in 100—200 ml water).

All the necessary precautions must be taken not to spread the infection. Intubation is done in the laboratory or in a separate room. The patient is protected by a mackintosh over which gauze moistened in a disinfectant is spread. On the table with the necessary instruments for intubation is placed a gauze steeped in disinfectant. The person performing the intubation must put on rubber gloves.

The bile is collected in a sterile container and 1 to 2 ml are introduced in test tubes with the two media, or seeded directly when intubation is done in the laboratory.

All the material used is autoclaved.

Collection of water

Water samples are collected in bottles sterilised in the autoclave (a teaspoonful of salt is introduced in the bottle before sterilisation — Barua, 1970 b).

After collecting, all the material used will be disinfected to avoid subsequent contamination of other water sources.

The water is brought to the laboratory as collected or may be immediately processed, either in a culture medium for cholera vibrios or by filtering it through a Seitz filter or through membranes of the Millipore type, when a larger amount of water will be examined.

If transport to the laboratory will take some time it is recommended to add 1 M NaOH, pH 9.2.

Water may be transformed into a culture medium by adding 1 : 10 of a concentrated peptone water made alkaline up to pH 9 with 1 M sodium hydroxide, in the presence of bromothymol blue as indicator (or pH paper indicator). By adding the necessary ingredients water may also be transformed into Gohar medium. It is then sent to the laboratory where the conventional processing starts.

Barua (1970 b) recommends the following procedure:

For collecting 200 ml water samples, 280 ml sterile bottles are sent containing 20 ml 10% peptone water with 5% NaCl.

Once the water bottle has reached the laboratory it is made alkaline (pH 9.2) and incubated overnight; 1 to 2 ml are then introduced into 10 ml alkaline peptone water and after 6 hours incubation the usual passages are performed.

The water may be filtered, as already mentioned, from the moment in which it is collected; asbestos (Seitz) or Millipore membranes are introduced into the collection flasks with conservation or enrichment medium.

Felsenfeld recommends the filtration of each litre collected in a sterile bottle containing sodium thiosulphate (1 g); the membrane is then introduced in a flask with a wide neck containing (50—100 ml) alkaline peptone water. Residual water is collected in a flask with a wide neck and if necessary is diluted with sterile saline solution or filtered through sterile gauze before being processed as are other water samples.

Collection of the material for isolation of the cholera vibrio from foodstuffs, fruit and vegetables

Food samples are collected and introduced into sterile flasks with or without conservation or enrichment media. The food samples may eventually be triturated and mixed with the medium.

Fluid food is processed as are the water samples or introduced directly into 1 to 3 ml amounts in the collection tubes with adequate media.

For fruit and vegetables there are two procedures:

(i) The fruit or vegetables are wiped with swabs moistened in sodium chloride solution and the swabs then introduced into one of the two media.

(ii) The fruit and vegetables are washed in sterile (boiled) water which is then treated as are the water samples or, when more concentrated, 1 ml is introduced in a sterile flask.

Pieces of clothes, linen or bed linen soiled by pathological products are introduced into a sterile flask containing one of the media mentioned. The pieces may also be washed and the washing water (boiled water, normal saline or conservative medium) introduced into adequate containers.

Transport of the samples to the laboratory

The samples are taken to the laboratory in compliance with all the necessary measures of precaution so as to avoid spread of the infection, but also to preserve viability of the germs.

When the samples are collected in hospital and processed in the hospital laboratory, the sealed containers are taken to the laboratory in metal or plastic boxes in case one of the containers might break or spill. The boxes may be disinfected or autoclaved.

The samples in the flasks must be surrounded by absorbant material (cotton wool, sawdust, etc.) and if several samples are taken each flask must be correspondingly marked so that it can be identified on an accompanying card.

Within a certain focus (closed community, family focus, etc.) metallic or plastic boxes may be used with several compartments for holding 10 to 12 flasks.

These boxes must be tightly closed with adhesive tape. They are put into another box and packed in paper (preferably parchment paper), tied with string and sealed with wax. The danger of contamination is mentioned on the parcel.

For water samples larger boxes are used and when several samples have to be sent isothermal boxes are used.

The material used for collecting the pathologic samples are disinfected or sterilised but when adequate conditions do not exist they are packed and sent together with the samples to the laboratory. This may include the metallic scoops, wooden or plastic material, special scoops incorporated in the stoppers, etc.

Disinfection of the containers with the faeces will be done with:

(i) lime chloride mixed in a proportion of 1 : 5 with the faeces and kept in contact for 2 hours, in a closed receptacle;

(ii) lime milk, double the volume of the stools, in the same way;

(iii) 1 : 5000 potassium permanganate in equal parts with the stool.

For collecting the samples only the flask for the sample will be taken into the room, the boxes and packing left outside.

After collecting the products the flasks are closed and wiped on the outside with a disinfectant (chlorigenic solution or sublimate); once outside the room they are introduced into boxes and sealed and the cards accompanying the samples are completed.

All the instructions must be strictly complied with especially when the laboratory is not within the precincts of the hospital. If it is, then the flasks with the samples may be introduced into a single container.

The products not processed when collected are transported to the laboratory on ice (in special thermos flasks, isothermal boxes or an improvised container); the samples in the Venkatraman and Ramakrishnan conservation medium may be transported as collected, irrespective of the temperature. For all products introduced in enrichment medium special precautions must be taken as the transport interval may be included in the incubation period of the culture.

All transport of pathological products to the laboratory must be made by special courier.

In the laboratory the parcels are unpacked with care and the packing burned or introduced into containers with disinfectants. Before opening the boxes they are wiped on the outside with disinfectant and after the flasks are taken out they are put in containers with disinfectants or autoclaved; the absorbant material is treated in the same way as the packing.

8. 2. 2. Bacteriological diagnosis

In principle, the product collected is introduced into the conservation medium (when the distance to the laboratory is greater or processing is delayed) or directly into the enrichment medium (when collected from in-hospital patients).

If the clinical and epidemiological data point to the possibility of cholera and a rapid diagnosis is necessary then the pathological product is brought to the laboratory, without being processed in order to establish the presence of the vibrio cholerae even before isolating it by the conventional methods. This includes:

(i) *gross examination of the stools* which may reveal aspects characteristic of cholera;

(ii) *examination of the fresh preparation* in order to reveal the presence of the mobile vibrios;

(iii) *the immobility test with specific sera* for identification of the vibrio in the fresh product;

(iv) *identification of the vibrios in samples stained with fluorescent antibodies*

The latter three tests are also used in all the stages of isolation and bacteriological identification of the cholera vibrio, thereby completing the classical methods of rapid techniques.

From the *conservation medium* passages are carried through *selective media*, either directly or after enrichment in adequate fluid media.

From the *enrichment medium*, seeded either directly with the product or from the conservation medium, the passages are carried through solid *selective media* for the isolation of colonies.

The suspect colonies isolated on selective media are identified serologically by slide agglutination and passed on adequate media for further identification tests. The pathological products must be concomitantly processed with a view to isolating other eventual pathogenic germs of the *Enterobacteriaceae* family that can produce cholera-like syndromes. This must be foreseen when collecting the samples, adequate conservation media being used, and in the laboratory where the pathologic product is also seeded on selective media for the *Salmonella* and *Shigella* group.

8. 2. 3. Conservation media

At present the Venkatraman and Ramakrishnan media are used almost exclusively for the conservation and transport of products collected from cholera patients. The original formula contains sea salt and boric acid; as there is no carbon source in its composition it permits conservation of the vibrio but not its multiplication.

The viability of the cholera vibrio in these two media is of up to 6 months whereas Salmonella and Shigella remain viable for only a few weeks. The great advantage of these media is that they can be kept, once seeded, at room temperature.

When these media are not available the samples may be conserved in 20% sodium chloride and made alkaline immediately after seeding with a drop of sodium hydroxide; this hypertonic saline as well as the normal saline or glycerolated saline solutions must be kept and transported in the cold.

The Cary-Blair medium has given good results (Zafari *et al.*, 1968), the vibrios living at least 7 days; it has, moreover, the advantage of being semi-solid.

For short conservation in the cold, Felsenfeld (1966) recommends alkaline peptone water or with 1 : 200 000 potassium tellurite (according to Gohar); kept one day at room temperature, however, it favours the abundant development of other bacteria which may cover *Vibrio cholerae*.

Sodium chloride with glycerine (30%) is used as conservation medium for germs of the *Salmonella* and *Shigella* group, as well as for enteropathogenic *E. coli*.

The conservation medium is distributed in 10 ml amounts in the collection flasks; the pathologic product is then introduced and mixed well with the medium before closing the flasks.

8. 2. 4. Enrichment media

Numerous enrichment media have been recommended, but according to the Indian physicians alkaline peptone water is just as efficient as other media and less complicated.

Potassium tellurite in an end dilution of 1 : 200 000 improves its efficiency. Potassium tellurite inhibits the supra-added microbial flora especially pyocyaneus, which in simple alkaline peptone water develops quickly and covers the growth of *Vibrio El Tor*; the staphylococcus alone proves resistant to the presence of tellurite.

Each tellurite lot should be previously controlled for establishing the optimal concentration (Barua, 1970 b).

Alkaline peptone water with tellurite has proved just as efficient as other enrichment media, but has a less complicated formula. Widely used is also peptone water with potassium tellurite and sodium taurocholate.

Barua (1970 b) also recommended peptone water with Teepol (0.1—0.2%), in which the sporulate germs and cocci are inhibited, but the vibrios grow more slowly. This medium appears to be better for transporting the samples.

As in the case of the conservation media, seeding is done on the enrichment media with 1 to 2 ml (1 g) amounts of the pathological product or and equivalent amount of the conservation medium in which the product was initially seeded.

The enrichment media are incubated at 37°C for several hours; the vibrio develops well within the first 4 to 6 hours. After this interval passages may be started on selective media, collecting with the loop from the surface of the medium or from the film that has formed.

When the products are badly contaminated and the enrichment medium is kept some time, the growth of cholera vibrios may be exceeded by that of the commensal flora. In such a case it is recommended after passages on selective media to seed a new enrichment medium, using 0.1—0.2 ml of the culture from the first tube, thus purifying the cholera vibrio culture. This procedure is frequently followed in the laboratories of Calcutta.

It is worthy of note that the enrichment medium indicated — alkaline peptone water with or without potassium tellurite or sodium taurocholate — promotes the selective growth of cholera vibrios only within the first hours of incubation; subsequently other intestinal bacteria rapidly develop (Salmonella, Shigella, coliform bacilli). Hence the passages in selective media must be done 4 to 6 hours after seeding the alkaline peptone water (including the duration of the transport when the product is collected directly on enrichment medium). Passages on enrichment medium should be performed every 8—10 hours; after this interval the associated flora (especially *E. coli)* multiply abundantly in spite of the alkaline pH.

8. 2. 5. Selective media

Almost all the selective media used at present for *Vibrio cholerae* owe their properties to bile salts (taurocholate, deoxycholate or total bile) and to potassium tellurite. The presence of tellurite or taurocholate transforms the alkaline agar into a highly selective medium. Some media also contain gelatin the characteristic aspect of which following its liquefaction, offers an additional criterion for recognising cholera vibrio colonies.

In order to differentiate the developing colonies, numerous selective media contain sugars, as a rule saccharose. The most frequently used selective media are:

(i) BSA (Bile, Salts, Agar);
(ii) TCBS (Eiken);
(iii) Monsur;
(iv) Aronson;
(v) Vibrio agar.

The Alkish medium, used in Turkey and based on the same principles is highly selective for cholera vibrios and strongly inhibits enterobacteriaceae except for Proteus. The Sakazachi medium is very selective but its formula is too complicated. *Vibrio agar medium* is considered more selective than BSA and less than TCBS.

Many of the selective media are supplied in the form of powders (TCBS-Eiken, Cholera Medium Oxoid) which increases the diagnostic possibilities of small laboratories.

The WHO recommends seeding of each sample on several selective media in order to increase the chances of isolating the vibrio. However, in typical cases, in the first or second day of the disease, it is not necessary to use very selective media since the stools represent almost pure cholera vibrio cultures.

The less selective media prove just as efficient: nutrient gelose and gelose-gelatin, pH 7.2, gelose with bile salts, pH 8.2, agar SS (Difco, BBL), pH 8, and the Istrati-Meitert medium in Romania. Feeley and Pittman (1963) did not obtain satisfactory results with the MacConkey medium.

In India, after much research work the BSA medium has been given preference, it is sufficiently selective and can be prepared with indigenous products at a low cost. The BSA medium is very dark, has the aspect of a simple gelose and is prepared with sodium taurocholate.

In BSA medium the cholera vibrio colonies develop within 10 to 14 hours. Their aspect is characteristic, as a rule round, slightly swollen, smooth and transparent, with bluish iridescence. The transparence of cholera vibrio colonies sharply differs from that of enterobacteriaceae colonies.

A skilled worker can readily recognise cholera vibrio colonies even when other bacteria develop; the Indian specialists consider that the cholera vibrio colonies may be recognised even in a proportion of 1 to 2 000 contaminating colonies.

Dr. Pal of New Delhi showed that recognition of cholera vibrio colonies developing on BSA medium can be improved by oblique illumination. However, this method does not always differentiate the cholera vibrios from *Pseudomonas*, *Commamonas* and NAG vibrios.

The "string" test may likewise be useful. It was modified by Neogy and Mukherji (1970) (cultivation on phosphate medium with 0.5% deoxycholate suspension and the formation of filaments) who do not, however, consider it specific of *Vibrio cholerae*.

The TCBS medium strongly inhibits the development of *E. coli*, but in Gangarosa's attempts at isolating the cholera vibrios it did not prove superior to the BSA medium, which has, moreover, the advantage of being simpler (Gangarosa *et al.*, 1966).

Following the experience acquired in numerous laboratories the Aronson medium has also been recommended: an alkaline gelose containing sachharose, dextrin and Schiff reagent as indicator. The cholera vibrio colonies usually develop in this medium at 10 hours after seeding; they are less transparent than those obtained in BSA and are at first pink, then after 18 to 24 hours they take on a bright red colour (as does *E. coli*) and grow in size.

In order to identify the suspect colonies, apart from the serological tests, fermentation is tested, using peptone water with sugars, or glucose broth in the hexamine test.

The *V. cholerae* biotype is established by the corresponding tests.

8.2.6. Culture media used for the bacteriological diagnosis of cholera

(1) *The Venkatraman and Ramakrishnan conservation medium*

Stock solution:

Boric acid	12.4 g
Potassium chloride	14.9 g
Distilled water	800 ml

After dissolving the ingredients the solution is cooled and brought up to a volume of 1 000 ml. To 250 ml of this solution are added:

Sodium hydroxide (5M)	133.5 ml
Sea salt	20 g

When sea salt is not available the following formula may be used:

Sodium chloride	27 g
Potassium chloride	1 g
Magnesium chloride ($MgCl_2 . 6H_2O$)	3 g
Magnesium sulphate ($MgSO_4 . 7H_2O$)	1.75 g

The volume is brought up to 1 000 ml with distilled water, the pH is adjusted to 9.2 and the solution filtered through filter paper.

The medium is sterilized in a globular flask (30 minutes at 110°C) and left to stand at room temperature until the next day, when the clear supernate is distributed in 10 ml amounts in collection containers; the filtrated solution can also be sterilised after being distributed into the containers.

(2) *Alkaline peptone water*

Peptone	10 g
Sodium chloride	10 g
Distilled water	1000 ml
pH 9.2	

After precipitation for 30 minutes at 120°C the medium is filtered and distributed in Jena tubes (180 × 18) or in MacCartney flasks in 10 to 15 ml amounts and sterilised for 20 minutes at 110 °C.

(3) *Alkaline peptone water with tellurite* contains in addition potassium tellurite in an end dilution of 1 : 200 000. It is prepared by adding 1% potassium tellurite solution (0.5 ml) to alkaline peptone water.

(4) *Gohar and Makrawi medium*

1 litre of medium contains:	
Peptone	5 g
Sodium carbonate	2 g
Sodium taurocholate	5 g
Potassium tellurite for an end dilution of 1 : 200 000	
pH 9	

(5) *Monsur medium*

Tryptic peptone	10 g
Sodium chloride	10 g
Sodium carbonate	1 g

The whole is dissolved in 1 000 ml distilled water, sterilised for 10 minutes at 121 °C. To this is added in sterile conditions:

sodium chlorate (10%)	30 ml
Potassium tellurite up to an end dilution of	1 : 200 000

For preparation of the solid medium agar is added in a proportion of 15 g per 1000 ml.

(6) *BSA medium* (Bile Salts Agar)

Peptone	10 g
Meat Extract (Lablemco)	10 g
Sodium chloride	5 g
Sodium taurocholate	5 g
Agar	20 g
Distilled water	1000 ml

The pH is adjusted to 8.8—9; the mixture is precipitated at 120 °C for 30 minutes, then filtered through cotton wool, and sterilised for 20 minutes at 115 °C. Poured into Petri dishes the BSA medium has the aspect of simple gelose.

The cholera vibrio colonies have a diameter of 1—2 mm, are smooth and very transparent in comparison to the other enteric bacteria that develop in this medium.

To the BSA medium lactose and bromothymol blue may be added as indicator, using 10% lactose solutions (5 ml) and a 0.2% alkaline bromothymol solution (4 ml).

When prepared the medium has a bright blue colour which is not modified by growth of the cholera vibrio or other lactose-negative bacteria but turns yellow when lactose-positive bacteria develop.

When the ingredients are not available the BSA medium may be prepared with other meat extracts and total bile salts. Poured into Petri dishes it can be kept for 4 to 5 days in the refrigerator.

(7) *Aronson medium*

Meat extract	10 g
Peptone	10 g
Sodium chloride	5 g
Agar	35 g
Running water	1000 ml

The mixture is boiled for 4 to 5 hours and heat maintained for sedimentation; the supernate is decanted and distributed in 100 ml amounts in 250 ml flasks.

To each 100 ml of boiling gelose 10% anhydrous sodium carbonate solution (6 ml) is added and boiled for 15 minutes then made up with:

Saccharose (20%)	5 ml
Dextrin (20%)	5 ml
Sodium sulphite 10%	2 ml
Fuchsin, saturated alcoholic solution	0.4 ml

The saccharose, dextrin and sodium sulphite solutions are previously sterilised separately, boiling for 30 minutes or by filtration.

The medium cooled to 40—45 °C is poured into Petri dishes and kept at +4 °C in the dark for several days. Each medium batch should be previously controlled or when used a collection *V. cholerae* culture should also be seeded.

(8) *TCBS medium* (Thiosulphate-Citrate-Bile-Succrose)

The TCBS medium is supplied in a dry form (Eiken) but may be prepared according to the following formula:

Eiken peptic peptone (Difco Bactopeptone)	10 g
Yeast extract	5 g

Total, dry bovine bile	5 g
Sodium cholate (Eiken)	3 g
Sodium thiosulphate	10 g
Sodium citrate	10 g
Iron citrate	1 g
Sodium chloride	20 g
Saccharose	20 g
Bromothymol blue	0.04 g
Thymol blue	0.04 g
Agar (Bacto)	15 g
Distilled water	1000 ml

After solving the ingredients, without the saccharose and indicators, the medium is sterilised for 30 minutes at 120° C then filtered through cotton wool while hot; the last two components are added and the whole sterilised for 20 minutes at 110° C then distributed into Petri dishes.

If the dry medium is used, 86 g are dissolved in 1000 ml distilled water and slightly heated up until the agar is completely dissolved, then poured into Petri dishes without previous sterilisation.

The greenish-blue medium turns yellow as soon as the cholera vibrio, which ferments saccharose, develops.

(9) *The Alkish medium*

Difco typtone	20 g
Nr. 3 Difco bile salts	10 g
Yeast extract	6 g
Sodium chloride	20 g
Sodium carbonate	3 g
Agar Noble	34 g

The ingredients are added in the order listed above and dissolved in the heat in 1700 ml distilled water, the pH being adjusted to 9.0—9.2. Then 5 ml of a 1% phenol red solution is added (50 mg) and the whole is sterilised for 15 minutes at 120 °C. After cooling 5 ml of a 1% potassium tellurite solution and 30 ml of a 30% saccharose solution are added under aseptic conditions.

(10) *Vibrio agar*

Yeast extract (Difco)	5 g
Trypticase (BBL)	4 g
Proteose peptone (Difco)	3 g
Sucrose	20 g
Sodium citrate ($2H_2O$)	10 g

Sodium thiosulphate ($5H_2O$)	6.5 g
Sodium deoxycholate	1 g
Sodium chloride	10 g
Ox bile	5 g
Sodium lauryl sulphate	0.2 g
Cresol red	0.02 g
Water blue	0.2 g
Agar	15 g
Distilled water	1000 ml
final pH 8.5	

The cholera vibrio colonies are bluish grey, translucent and seem to be larger than that of other bacteria that eventually show up; in contrast to the colonies grown on TCBS medium those isolated on Vibrio agar are readily agglutinated by the specific sera.

The *Proteus* and *Aeromonas* colonies are smaller and bluer, and the *Klebsiella* are larger, opaque and mucoid.

The authors recommend the concomitant use of Vibrio agar with the TCBS medium.

(11) *Medium for hexamine test*

(a) The first solution is prepared with:

Crystal hexamine	1 g
Nutrient broth pH 7	10 g

The solution is sterilised by filtering through a Seitz filter then kept in the refrigerator.

(b) The second solution includes:

Hexamine broth (solution a)	12.5 ml
Sterile glucose broth (1%)	87.5 ml
Bromocresol purple (0.4%)	4.5 ml

The solution is distributed in 1 ml amounts in haemolysis test tubes.

The bromocresol solution is prepared by triturating 1 g in 5M sodium hydroxide (9.25 ml), then adding distilled water up to a volume of 250 ml.

Apart from the media that favour particularly the isolation and identification of the cholera vibrio, lactose gelose with bromothymol blue or the Istrati-Meitert medium with dry bile can also be used, especially for passage of the suspect colonies that have grown on selective media (BSA, Aronson, TCBS, etc.).

For the biochemical tests as well as for isolation and conservation of the isolated strains the conventional media for this purpose are used.

(12) *Other media*

Apart from the media for the isolation and identification of the cholera vibrios, the pathological products — the stools — are seeded on the media recommended for the bacteriological diagnosis of enterobacteriaceae:

(i) conservation media (30% glycerinate solution);

(ii) enrichment media (with selenite, Müller-Kaufmann, etc.);

(iii) selective and differential media (Wilson-Blair, MacConkey, Leifson Istrati-Meitert, etc.).

8. 2. 7. The stages of the bacteriological diagnosis

8.2.7.1. *Examination of the stools*

The stools collected from the patients are examined and seeded for isolation of the cholera vibrio; parallel bacteriological investigations are carried out in order to determine whether the infection is caused by a bacterium belonging to the Enterobacteriaceae (especially *Salmonella* and *Shigella*).

Gross examination of the stools may reveal characteristic elements; watery stools containing colourless or whitish rice-like grains. The presence of red blood cells may modify the aspect, giving it a reddish tint.

However, the rice water aspect is more seldom encountered nowadays. Even in cases of bacteriologically confirmed cholera the stools may be dysenteriform or even well formed faeces.

Examination of the fresh sample on slides covered with a coverslip is a rapid procedure for the diagnosis of cholera. In about 80% of the cases examined at the onset of the disease the vibrios with their characteristic mobility are readily recognised. The slides can be examined in the light microscope, and by the phase contrast and dark field methods (recommended by the Indian specialists). No special processing is necessary.

In the fresh sample the vibrios are identified by the *immobility test* recommended by Benenson and his co-workers (1964). The test is performed as follows:

Two drops of fluid stool are put on the same slide: one suspended in 0.9% sodium chloride, and the second from the same stool is suspended in cholera antiserum *(OI, A)*, not diluted or diluted at a titre indicated on the vial for slide agglutination. A cover slip is then put over the slide which is examined in the microscope, preferably by the dark field method.

The cholera vibrios are very mobile in the saline suspension and lose their motility in the cholera antiserum, forming clusters; this may actually be considered a slide agglutination test.

Very good results are obtained with this test when the antiserum is used in very high titres.

In the immobility test it is very important not to use sera containing stabilising substances (for instance sodium methiolate) since, owing to their chemical action upon the vibrios, all the results may be positive.

When the stools are harder the sample must be suspended 1 : 10 in 0.9% sodium chloride but the results are not so good.

The immobility test can also be carried out with the seeded conservation medium, but the interpretation is more difficult as the sample is already diluted.

The test can also be performed after 6 to 8 hours enrichment in alkaline peptone water, using a droplet from the surface of the medium. This raises the proportion of positive results up to 80%.

The specialists in India consider the immobility test on slides as very useful. It is currently used in cholera hospitals and sometimes performed at the patient's bedside, using a portable microscope. Positive results are obtained as a rule at a concentration of 10^4-10^5 bacteria ml^{-1} the test being therefore more sensitive than immunofluorescence.

The sample stained with *fluorescent antibodies* is likewise a rapid and sufficiently specific diagnostic method; its degree of sensitivity is appraisable but Indian investigators consider that positive results can only be obtained at a concentration of 10^6-10^9 cholera vibrios per gram stool. Unspecific fluorescence caused either by the associated microbial flora or amorphous particles of vegetable origin, may falsify the results, especially when the concentration of the vibrios is lower than the one mentioned. In the patent cholera cases in which the stools represent a pure cholera vibrio culture the concentration is much greater and immunofluorescence is then useful. On the other hand, for the control of carriers, contacts or convalescents the method is not sufficiently sensitive.

For immunofluorescence fluid stools are used as collected or, if they are harder, they are suspended 1 : 10 in 0.9% sodium chloride; 15 minutes later when large particles separate, a drop is put on the slide, then dried at room temperature and fixed with acetone (10 minutes) or over a flame.

Staining is done in a moist chamber (Petri dishes with cotton wool or moist filter paper), maintaining the smears covered with fluorescent cholera antiserum for 20 minutes at 37°C; the slides are then washed for 10 minutes in a saline solution, pH 7.2, or tap water, and dried in the air or with filter paper.

With the indirect method the smears are covered with cholera antiserum (20 minutes at 37 °C in a moist chamber), washed (buffered saline for 10 minutes) and then treated with fluorescent globulin antiserum to the species that supplied the cholera antiserum used in the first staining stage. Washed and dried the smears are examined using buffered gylcerin for the immersion.

The staining technique is chosen according to the available sera. The dilution of the fluorescent sera is that indicated on the vials; as a rule the globulin antiserum is diluted 1 : 4.

The morphological aspects vary very much since in the patient's intestine the cholera vibrios range from coccoid to filamentous, and it is often difficult to differentiate them from other abundant fluorescent formations in the stools. Only when the samples are very rich in vibrios can they be identified by their typical morphology and specific staining.

Nestorescu *et al.* (1965) insisted that differentiation demands a concentration of at least 10^9 g^{-1} sample.

With qualitative improvement of the immune sera the method will play an important role in the orientative diagnosis of the first cases, in field conditions.

Immunofluorescence is at any rate very useful in screening samples in a declared focus, especially when there is a high concentration of bacteria in the faeces.

The diagnostic results are greatly improved when immunofluorescence is combined with cultivation. After 3 to 69 hours incubation the material collected from the surface of the seeded enrichment medium is used for smears, the unspecifically stained elements usually encountered in the faeces being absent.

However, up to date, immunofluorescence is not widely used due to the unspecific fluorescence of the smears prepared directly from the patient's stool. Once the antibiotic treatment is instituted the number of vibrios in the stools decrease sharply. As with the carriers, few bacteria are eliminated and immunofluorescence is no longer efficient without previous enrichment of the sample in alkaline peptone water.

•

Apart from rapid orientative examinations the usual bacteriological tests for the isolation and identification of the cholera vibrio are carried out, also with a view to isolating other bacteria that might have caused the disease.

(a) When the faeces are brought directly to the laboratory they are seeded in an enrichment medium (alkaline peptone water, preferably with tellurite) and on enrichment media for different enterobacteriaceae (for instance the selenite medium, Müller-Kaufmann medium). Direct seeding may also be done on selective media for cholera and for Enterobacteriaceae (MacConkey, Wilson-Blair or Istrati-Meitert medium).

The WHO experts recommend seeding on selective media in Petri dishes especially when collecting is done with swabs that are introduced into the enrichment medium only after seeding.

When the sample is brought in a conservation medium (Venkatraman and Ramakrishnan) the samples are taken up with a loop and passed on selective media, concomitantly seeding enrichment media.

When the sample is seeded on alkaline peptone water before reaching the laboratory then incubation is continued in a thermostat.

(b) Passages on selective media are only carried out after the alkaline peptone water has been kept in the thermostat for 4 hours. The operation is repeated every two hours. When there are indications (in the focus or other epidemiological data) that the vibrio is of the El Tor type incubation may take longer for the passages on selective media to become positive.

Seeding from the alkaline peptone water may be done in a new tube with the same medium. This second enrichment purifies the commensal bacteria. Barua, 1970 b, considered that for the second enrichment the broth-trypticase taurocholate-tellurite medium was better than peptone water.

After a short incubation in the enrichment medium the upper part of the culture, eventually the film that has formed, is used for direct examination between slide and coverslip, completed by the imobility test and immunofluorescence.

Seedings are also done from the alkaline peptone water on the currently used media in the laboratory for the isolation of intestinal bacteria.

(c) Finally, the selective media for cholera vibrios are seeded with the product as such or with the media in which it was brought to the laboratory (conservation medium or alkaline peptone water) or after successive use of the latter.

The enrichment media appear to be particulary adequate for the products from convalescents; within the first two days of the disease seeding on selective media is preferable, but both are used in routine examinations (Barua, 1970 b).

Felsenfeld recommends that seeding should be done on at least two types of media: one with a strong inhibitory potential (TCBS, Monsur, Cholera Medium Oxoid, etc.) and the other with less inhibitory properties (alkaline agar).

The selective media are incubated at 37 °C for at least 10 to 12 hours; the dishes are then controlled periodically for detecting the colonies with a characteristic aspect.

The cholera vibrio colonies are identified by their smooth aspect and particular transparency and, obviously, in terms of fermentation of the sugars in the medium.

As already mentioned, the cholera vibrio colonies may also be identified by examination with the binocular magnifying glass at an oblique illumination. Felsenfeld (1966) draws attention, however, to the fact that the examiner must be used not only to the smooth colonies, that are more frequently encountered but also to the R (rough) ones. The light beam, passing through a blue filter in daylight is directed upwards with the aid of a concave mirror through the bottom of the Petri dish at an angle of approximately 40 °C.

The cholera vibrio colonies appear to be circular, swollen, smooth, finely granulated, translucent, with irisation and a tendency to irregular concentric striations. Although this aspect may vary in terms of the age of the culture, it permits differentiation of other bacteria that give rise to more or less transparent colonies with a different aspect.

The suspect colonies are identified and the vibrio type established.

8.2.7.2. *Identification of the cholera vibrios*

The colonies are intially identified by *slide agglutination* with a polyvalent cholera antiserum, diluted with 0.25—0.50% phenol saline solution in the titre indicated on the label.

The WHO experts recommend examination of the suspect colonies identified in selective media, but agglutination may be very well performed also on BSA. When agglutination with the polyvalent antiserum is positive then agglutination is continued with the specific Ogawa and Inaba immune sera.

If a suspect colony does not react with the polyvalent cholera antiserum, at least another 5 to 10 colonies should be examined. Finally, agglutination may also be tried with a sample taken with a loop from the mass of the culture, since agglutinable with non-agglutinable vibrios may coexist.

The rest of the colony that have a positive agglutination is then passaged on inclined gelose, pH 7.2—7.6, and on a differential medium (lactose gelose with bromothymol and 1% thiosulphate) Istrati-Meitert medium or — as recommended by Felsenfeld and the WHO — on a ferrous sulphate Klieger or TSI medium. Concomitant seeding is recommended in heart and brain broth and in peptone water, since this curtails the duration of the identification tests.

The cultures obtained in these media serve for a renewed confirmation of the identity by both agglutination on slides and in tubes and by biochemical tests. These cultures are also used for microscopic examination (stained smears, freshly prepared for establishing mobility and its inhibition).

For *agglutination in tubes* the following method is used:

In haemolysis tubes, progressive dilutions of the polyvalent cholera antiserum are prepeared and / or of the monovalent Ogawa and Inaba antisera up to the titre indicated; the dilutions in a 1 ml volume are prepared with normal saline.

From the fresh culture on inclined gelose a thick suspension is prepared in normal saline (for the sake of safety the normal saline is treated with 2% phenol suspended in 2—3 ml tubes) so as to obtain a concentration of 2×10^9 germs ml^{-1}.

To each tube with serum dilutions a drop of suspension is added, using as control a tube with normal saline.

The tubes are incubated for 2 hours in a water bath at 52 °C and then, after the first reading, are left at room temperature until the next day when the definitive reading is done.

The result is considered positive when visible agglutination occurs (++ or +++) up to at least half the serum titre.

A positive slide agglutination or tube test must be confirmed by biochemical tests since unspecific agglutination may take place with other intestinal microbial species.

Tube agglutination appears to be necessary for the first strains isolated in a focus; subsequently, slide agglutination is sufficient for recognising the cholera aetiology.

The identification tests based on vibriolysis (the Pfeiffer-Isaeff phenomenon) recommended in the textbooks is more difficult and not essential for diagnosis in the field.

Biochemical identification is carried out according to the Heiberg fermentation scheme; the sugars used in the tests are saccharose, mannose and arabinose in a final concentration of 0.5% in peptone water, with bromothymol blue as indicator.

Both the classical and the El Tor vibrios attack saccharose and mannose but do not act upon arabinose, and therefore form part of the first group on Heiberg's scheme.

Several tests have been used to differentiate the cholera vibrio from other enterobacteriaceae:

(i) The *indol test:* Kovacs reagent (1—2 ml) (*p*-dimethylaminobenz-aldehyde, (5 g) amyl alcohol (75 ml) and concentrated hydrochloric acid 25 ml) is added to a cholera vibrio culture in broth; after stirring the result is read in 5 minutes.

(ii) The *hydrogen sulphide* test is assessed according to the change that takes place in the medium with ferrous sulphate (Klieger or TSI medium) or darkening of the filter paper bands impregnated with lead diacetate $Pb(OOCCH_3)_2$ introduced in the tubes with seeded broth.

(iii) *Lactose fermentation* is performed in tubes or in differential media containing this sugar.

(iv) The *breakdown of urea* is determined by seeding on media containing this substance.

(v) The *acetylmethylcarbinol test* — the Voges-Proskauer reaction — — is best carried out by the Barrit technique. To 1 ml 48 hour cultures in Clark medium 0.6 ml of a 5% alcoholic solution of α-naphthol is added and then 0.2 ml of a 40% potassium hydroxide solution; the mixture is stirred and the results read after 1 to 4 hours or even sooner if the tube is introduced in the thermostat.

(vi) The presence of *oxidases* is revealed by one of the following techniques:

(a) On the colonies developing in the culture medium a drop of a tetramethyl-*p*-phenylene diamine chloride (1%) is put (not more as the reactant must not be in excess). The result is positive only if the blue tint appears within 20 to 30 seconds. The test cannot be performed on media containing tellurite. Oxidase positive colonies can be carried through further passages only within at most one hour as they lose their viability after this interval.

(b) The presence of oxidases may also be revealed by filter paper strips moistened in the reagent and dried in the thermostat. They may be kept at room temperature and in the dark for several months.

For use, the strips are moistened in normal saline and put on a slide; a sample from a fresh culture on inclined gelose is taken with a loop and added to the filter paper. When positive, the filter paper turns blue within 20 to 30 seconds. Any change after this interval should not be taken into consideration. Positive and negative controls must be used in these tests.

The isolated germs are identified by the following tests:

Determination of the V. cholerae type is done by testing

(i) the sensitivity of the vibrio to Polymixin B;

(ii) its haemagglutination potential in chick red blood cells;

(iii) its haemolytic potential on sheep red blood cells;

(iv) behaviour in the medium with hexamine;

(v) sensitivity to group IV bacteriophages (and at present also to group V bacteriophages).

Determination of the sensitivity of the cholera vibrio to Polymixin B is performed in gelose to which Polymixin B in a final concentration of 15 $\mu g\ ml^{-1}$ is added. The tested strain is previously seeded in broth, preferably with heart extract.

The gelose with Polymixin B is seeded in spots with a young 2 hour culture. This technique is used especially when many samples have to be tested as at least six can be tested in the same Petri dish.

The sensitivity to Polymixin B is also tested by the diffusiometric method, using discs or tablets containing 50 μg Polymixin. Discs with cholistinsulphate can likewise be used.

The seeded medium is kept in the thermostat over night. At the Polymixin B concentration referred to the El Tor vibrio develops whereas the classical strain is inhibited.

Haemagglutination of the cholera vibrios is determined on a slide with a drop of a thick suspension of cholera vibrio cultivated on gelose mixed with a drop of a 2% suspension of chick red blood cells, previously washed thrice in normal saline.

The El Tor vibrio haemagglutinates the red blood cells within 10 minutes; haemagglutination does not take place in the presence of the classical vibrio.

Haemolytic potential of the cholera vibrio on sheep red blood cells is tested with a 24 hour culture in broth containing heart extract, pH 7.4. Felsenfeld recommends that this medium should also contain glycerine (1%) in order to reveal the haemolytic potential of the El Tor strains. However, in some instances cultures in peptone water are used.

To 0.5 ml of this culture 0.5 ml sheep red blood cell suspension is added (washed and adjusted to a concentration of 3%). The samples are kept for two hours at 35—37 °C and then overnight at +4 °C; the results are read the next day.

The controls of the reaction are:

(i) red blood cells in a concentration of 3% (0.5 ml) in normal saline (0.5 ml) used as control for the red blood cells;

(ii) red blood cells in a concentration of 3% (0.5 ml) in 0.5 ml broth with unseeded heart extract, used as control for the cultivation medium.

The haemolytic activity of the cholera vibrios is also determined on gelose containing 5% blood.

The El Tor cholera vibrio produces haemolysis of sheep red blood cells whereas the classical vibrio does not.

Behaviour of the cholera vibrio in hexamine medium is tested by seeding the hexamine medium with a 3 mm loop, distributed in 1 ml amounts in haemolysis tubes with a 24 hour culture (broth) of the vibrio to be tested (Koh *et al.*, 1967).

The seeded medium is maintained for 24 hours at 37 °C; in the presence of the classical strain (which does not develop) the medium remains unchanged, whereas following multiplication of the El Tor strain the purple medium becomes turbid then bluish-grey.

Testing the sensitivity of cholera vibrios with group IV and group V bacteriophages. The sensitivity of the cholera vibrio strains is determined using a 2 hour culture in 1% peptone water (pH 7.2). This culture is seeded with the loop over areas measuring 1 cm diameter in Petri dishes with 2% gelose (pH 7.2). After drying, a drop of group IV or group V bacteriophage suspension is added over the seeded areas. The dishes are incubated overnight at 37 °C.

The classical vibrio is sensitive to bacteriophage group IV whereas the El Tor strain is resistant; the behaviour of the two strains to group V bacteriophages — recently introduced by Mukerjee — is just the opposite.

Several other tests have been recommended for differentiating the cholera vibrio types: the Voges-Proskauer test, thermostability at 56 °C (Gispen test), agglutination with serum and sodium carbonate, precipitation in the presence of mercury chloride (Tanamal test), and others usually performed in specialised laboratories.

•

We have shown the various stages of the bacteriological diagnosis as practiced in the most frequently encountered situations.

In the case of carriers the general scheme is the same, with an earlier administration of a saline purge or magnesium sulphate respectively, as already mentioned. In one of the methods recommended, faeces (3 g) in alkaline peptone water (50—100 ml) is used.

As the vibrios are uniformly distributed on the surface and in the faeces the samples obtained following spontaneous excretion of the stools or by intubation are fragmented in alkaline peptone water (10 ml). As the samples from carriers contain a smaller amount of vibrios (10^3-10^5 g^{-1}) and might be masked by the commensal flora, it is recommended to repeat the passage through alkaline peptone water.

•

The following are some of the working schemes more frequently applied in India and recommended by the specialist literature.

(a) Patients (Madras Hospital)

(i) seeding of the stools on alkaline peptone water;

(ii) the seeded tubes are sent to the laboratory (King Institute);

(iii) after 6 hours incubation the samples are seeded on Aronson and/or BSA media;

(iv) slide agglutination and the other tests for the identification of suspect colonies are performed.

(b) Suspects and carriers in the field (Madras)

(i) the faeces are collected with a swab and introduced in a test tube with alkaline gelose, then sent to the laboratory (King Institute);

(ii) the alkaline peptone water is seeded with the swab;

(iii) incubation for 6 hours and passage in Aronson and or BSA medium;

(iv) agglutination and isolation for complete identification of the suspect colonies.

(c) *Carriers (Calcutta)*

(i) collection of the faeces in a flask with peptone water;

(ii) incubation for 6 hours, after which seeding in Petri dishes is performed; the rest of the contents in the flask is introduced into a globular flask with alkaline peptone water (150 ml);

(iii) slide agglutination and other identification tests of the suspect colonies, in Petri dishes;

(iv) passage in Petri dishes (from the content of the globular flask with peptone water) and identification of the suspect colonies.

(d) *From latrines (Calcutta)*

(i) collecting on Venkatraman and Ramakrishnan medium;

(ii) passage in Petri dishes with BSA medium;

(iii) the content of the 10–20 ml flask is introduced into a globular flask with alkaline peptone water (150 ml);

(iv) incubation for 6 hours and passage on BSA medium.

(e) *Pierce and his co-workers* (1970 a) used the following working scheme:

(i) samples were collected from the spontaneous stools excreted by the patient on admission, and afterwards swabs introduced into the rectum;

(ii) the fluid stool samples are seeded on selective media for cholera vibrio (BSA, sometimes TCBS), on agar with meat extract, on SS agar or deoxycholate agar;

(iii) the swabs are introduced into alkaline peptone water (pH 9.2) and after 6 hours this medium is seeded on BSA and nutrient agar.

(f) *Benenson and his co-workers* (1964) worked according to the following scheme:

(i) the samples are collected with a swab;

(ii) the content of the swab is immediately expressed on agar gelatin and agar with tellurite-taurocholate gelatin;

(iii) after seeding on selective media, the swab is introduced in peptone broth with bile and tellurite and incubated over night; next day passages are carried through on agar gelatin and agar with tellurite – taucocholate-gelatin.

(g) The WHO (1970) applied the following scheme for the detection of carriers:

(i) the samples are collected with a swab;

(ii) seeding is done an alkaline peptone water and incubated at 37 °C for 5 to 6 hours;

(iii) the passages are then performed with a loop on BSA and TCBS media; from this medium 0.1—0.2 ml is taken up with a pipette and again introduced into a tube with alkaline peptone water (5 ml), which is incubated for 5—6 hours.

(iv) from the new cultures BSA and TCBS media are seeded.

As may be seen these are but different variants of the same classical scheme.

Fluid stools or swab → conservation medium → Selective media
Fluid stools or swab → conservation medium → enrichment medium → Selective media
Fluid stools or swab → enrichment medium → Selective media
Fluid stools or swab → Selective media

using one or several of the media recommended.

At any rate, parallel to the tests for isolation of the cholera vibrio the current tests must be carried out for establishing an eventual "banal" aetiology of the disease (germs of the Enterobacteriaceae family) or the association of such an infection with cholera; in Madras we had the occasion to observe two cholera cases associated with typhoid fever.

Evidence of the *kappa bacteriophage* in the stools of El Tor vibrio carriers is obtained by the technique of Takeya *et al.* (1967 a and b), i.e. multiplication in the presence of the indicator strain.

A young culture of *V. cholerae*, strain H 218 Sm^r is diluted 1:100 in broth, pH 8.2 in the presence of streptomycin (300 $\mu g\ ml^{-1}$). To 5 ml of this suspension 1 ml fluid stool is added; when the stool is harder the suspension is prepared from faeces (0.5 ml) in (1 ml), broth pH 8.2.

After 7 and 24 hours incubation respectively at 37°C the supernate of the centrifuged culture is tested in spots on the indicator culture (H 218 Sm^r) grown on gelose in Petri dishes.

The bacteriophage may be isolated from 90% of the carriers. The test may also be applied to patients who no longer excrete vibrios in their stools following the antibiotic treatment.

8.2.7.3. *Bacteriological examination of other products*

When other products are taken for isolation and identification of the cholera vibrio (vomit, intestinal content collected from cadavers, fragments of clothes or linen, the water with which fruit or vegetables have been washed) the bacteriological investigation is similar to that of the faeces.

Water is examined according to the scheme presented both in the case of seeding from the filters through which a large amount of water has passed or of samples transformed into culture media by adding concentrated alkal ne peptone water. Samples from this water are seeded on selective media and rapid identification methods are applied – the immobility test and immunofluorescence.

When the latter test is applied the water sample may be centrifuged for a long time; smears are prepared from the sediment and stained by one of the techniques mentioned.

We likewise recommend the technique described by Popescu and Georgescu (1965): the water samples are filtered through nitrocellulose membranes with a pore dimension of 3 μm, mounted on a Seitz filter, with a 3 cm diameter; water (100 ml) is passed through the filter. After filtration the membranes are dissolved in acetone (5 ml); the suspension obtained is centrifuged (10 min. at 3000 r.p.m.). The sediment is again washed in acetone (5 ml) and centrifuged. Smears are then prepared from droplets, dried in the air and fixed over a flame.

Immunofluorescent staining of the smears is done by the known techniques. Evidence of the microbial bodies may be found under these conditions at a concentration of at least 10 000 bacteria ml^{-1}.

8. 2. 8. Serological tests

The serological tests are used for establishing a retrospective diagnosis. For this purpose paired sera are used, the first collected 48 hours after the onset of the disease and the second between day 7 and 15.

Serological survey is an efficient, easy method for recording chronic carriers, the agglutinating or vibrocide antibody titre being very high: 1 : 641 up to 1 : 1280. As these titres generally fall sharply 6 to 12 weeks after recovery, when they are still high 12 weeks or more after the acute episode, a detailed bacteriological examination becomes necessary.

Serological tests are often used in surveys in the field in order to assess the results of cholera vaccination.

The immunologic response is evaluated by titration of the agglutination, neutralisation and vibriocide antibodies. Recently, Ghosh *et al.* (1969) proposed the use of a simple technique in which red blood cells sensitised with soluble cholera vibrio antigens are used; passive haemagglutination has the technical advantage of necessitating only a drop of blood collected from the pulp of the finger.

8. 2. 9. General considerations on the laboratory diagnosis of cholera

Even if not indispensable for the treatment of patients a rapid laboratory diagnosis of all cholera cases is of importance for organising the prevention and control of the disease.

The laboratory diagnosis in cholera is essentially bacteriological, based on isolation and identification of the pathogenic agent, chiefly found in the patients' stools.

Bacteriological investigations are of particular importance for detecting cholera vibrio carriers.

The samples must be examined immediately after collecting and when this is not possible – when a round-the-clock laboratory does not exist—then the samples must be kept in the refrigerator.

Convenient media must be chosen for the bacteriological examination, with various degrees of selectivity if possible. Moreover, the laboratory is obliged to carry out investigations on the possible presence of other enteropathogenic bacteria, applying the known classical methods. Apart from the schemes presented direct, rapid seeding of selective media is also recommended.

Repeated passages on enrichment media raises the proportion of positive results by at least 10%.

The incubation must be prolonged in the case of El Tor vibrios.

The rapid identification techniques (immobility test with freshly prepared specific serum and immunofluorescence) may be performed at any of the stages of the bacteriological examination.

8. 3. Organisation of the laboratory diagnostic activity

The importance of the diagnosis in cholera demands a rigorous organisation of the various stages, establishing the level at which the various tests and examinations are performed and the epidemiological measures to be taken in case of a positive diagnosis.

As the implications of a diagnosis of cholera in the life of the community are well known it is necessary to establish previously the way in which the aetiology of the first cases will be determined.

An initially false diagnosis in smallpox, with all its consequences, is preferable to a falsely negative one in view of the extreme gravity of the disease but this is not admissible in cholera where the diagnosis can be correctly established within 24 hours or less.

The WHO considers that in the organisation of research services local laboratories should be created and a central reference laboratory.

The local (district or hospital) laboratory must be able to carry out bacteriological investigations for determining the presence of cholera vibrios and other enteropathogenic bacteria up to their seeding in Petri dishes with selective media and identification of the suspect colonies by slide agglutination.

The isolates from every patient must then be sent to the central laboratory for confirmation. To be able to cope with this responsibility the local laboratory must be equipped with the necessary material and have skilled personnel, able to put at the disposal of the health authorities the necessary data for making a decision.

It is recommended to examine the sample at the patient's bedside with a portable microscope and perform direct seeding on selective media.

The central reference laboratory organised on a national scale has multiple tasks:

(i) to establish the techniques and furnish the necessary materials to the local laboratories;

(ii) to examine the products received from the local laboratories;

(iii) to help the local laboratories by adequate instruction of the personnel and the dispatch of a specialised staff;

(iv) control and maintenance of the competence of the personnel in these laboratories.

The central laboratory must have at its disposal complete equipment for the diagnosis.

The central laboratory also represents a centre of production and research and is responsible for investigating the eventually unidentified cases and the carriers.

The WHO coordinates the activity of the national centres by means of an international reference centre.

Special laboratories must be prepared for performing the necessary examinations for a diagnosis of cholera; once a cholera focus has been declared the hospital in which cholera patients and suspects are admitted must also be provided with a laboratory for the diagnosis of enteric infections (cholera, Salmonella or Shigella infections).

The main duties of this laboratory are, in principle:

(i) to prepare the material used for collecting the pathological products from the patients and suspects;

(ii) to collect the samples for the bacteriologic examination;

(iii) to establish the bacteriological diagnosis by isolation of the cholera vibrio or other pathogenic germs that might have caused the disease.

The products are examined in the local laboratory by direct tests (fresh product, immobility test, immunofluorescence) and successive cultures for identification of the suspect colonies; the isolated strains are then sent to the central laboratory (in Romania the Cantacuzino Institute).

The latter confirms the presumptive diagnosis of the local laboratory either by direct examination of the samples (when two samples are collected and one sent directly to the central laboratory) or by controlling the isolated strain seeded on inclined gelose. Specialists may also be sent to the local laboratory for confirming the diagnosis.

When the diagnosis of cholera is confirmed, then the district or hospital laboratory must:

(i) establish a presumptive diagnosis of the suspect cases admitted to separate wards;

(ii) prepare the material that will be used for collecting the products in the field and transporting them;

(iii) examine the samples collected in order to establish the sources of infection, contamination of the environment and the carriers from among the contacts;

(iv) to isolate and identify the cholera vibrios of the patients admitted to hospital so as to insure that no convalescent has become a carrier.

Hence, the hospital laboratory advances with the diagnosis up to isolation of the suspect colonies and serologic identification by slide agglutination. If necessary, the vibrio type may be determined by haemagglutination of chick red blood cells, haemolysis of sheep red blood cells, sensitivity to Polymixin B and hexamine.

In Romania, lysotyping is done by the Cholera Unit of the Cantacuzino Hospital. All isolated strains are kept by seeding on gelose.

Of particular importance is the control at the frontier of people coming from countries with a notified cholera focus. The health officers at the customs will have the material necessary for collecting samples (containers, flasks, conservation medium, purges, flasks), disinfectants, autoclave, etc.

•

The activity of a laboratory for the diagnosis of cholera can only be efficient if carefully prepared beforehand:

(i) a special section or unit is organised in the central laboratory upon which the affected area depends;

(ii) the personnel must be carefully instructed and trained in the methods of diagnosing cholera, eventually serving in a laboratory abroad where this diagnosis is currently performed;

(iii) detailed written instructions must be drawn up referring to the bacteriological diagnosis and use of the materials and equipment of the laboratories, insisting especially upon the specific methods that permit a rapid diagnosis;

(iv) the diagnostic methods and techniques must be investigated and improved upon, in keeping with the progress made in other countries;

(v) periodical refresher courses, with practice in situations as close as possible to reality;

(vi) preparation and renewal of the stock material for the diagnosis (agglutinating cholera antisera, labelled sera for immuno-fluorescence, dry culture media, bacteriophage kits, etc.).

The laboratories must be organised so as to avoid any possible risk of infection either to the personnel or the environment. This also involves protection equipment, the possibility of disinfecting residual waters, burning the contaminated material that cannot be disinfected or sterilised, etc.

In principle, a laboratory for the diagnosis of cholera must have two rooms and an entry with a special cubicle for undressing and changing shoes. Mats steeped in disinfectants are put at the entrance to the laboratory. The tables are covered with gauze moistened in a disinfectant. The samples brought to the laboratory are unpacked on the tables thus protected and the packing introduced in buckets with disinfectants or in the container with infected objects to be destroyed.

The samples will be processed taking all the measures of precaution specific of a bacteriology laboratory.

The protection garments are boiled and the instruments autoclaved in the laboratory. No food or smoking is allowed in the laboratory.

Chlorigenic substances are recommended as disinfectants. No material can leave the laboratory without being sterilised.

The personnel of the laboratories must be vaccinated and revaccinated against cholera; during the alert period vaccination is repeated every three months. When an accidental contamination is suspected antibiotics (tetracyclin, chloramphenicol) will be immediately administered in doses of 500 mg, every 6 hours for at least 2 days.

9. Epidemiology

9. 1. Source of infection

In cholera the only reservoir of infection is the patient and the carrier. No animal is known up to date to be receptive under natural conditions to cholera; the animals in the contaminated areas will harbour and excrete vibrios passively, but do not become a reservoir of infection.

The patient is contagious from the very first signs of the disease (according to some authors during the last days of incubation). He is contagious as long as he eliminates bacteria in the stools, that is during the duration of the disease, convalescence and sometimes even after clinical recovery. In convalescence elimination lasts in the mean 15 days after the onset, as established by various investigations carried out in the course of cholera epidemics. In the 1944 cholera epidemic of Egypt 96% of the 75 cholera convalescents excreted bacteria for 12 days after the onset and 4% for 20 days. Similarly during the 1947 epidemic in Egypt in a study carried out on 250 cases it was found that 86% of the convalescents no longer excreted germs after the 15th day. On the 21st day 96% of the convalescents no longer excreted bacteria. Finally, according to other authors 80% of the convalescents are negative after day 14 and after 30 days there are only 1% excretors.

The El Tor cholera convalescents studied in the 1962 epidemic in Taiwan revealed absence of the vibrio in 74% of 279 patients tested on the fourth day of the disease; this proportion increased to 98% on the ninth day and 100% on the 14th day after the onset.

The Indian physicians and researchers have shown that following the treatment with antibiotics the duration of excretion in convalescence is much shorter. In the Hospital for Communicable Disease in Calcutta 90% of the patients became negative after 1 to 2 days treatment, and in most cases excretion of the bacteria never went beyond 15 days even in the untreated cases.

The Institute of Hygiene and Public Health of Calcutta, following the investigations conducted in this connection, showed that an excretion of cholera vibrios for more than 45 days is an exception.

However, these data must be reconsidered in view of the findings made subsequently in the Philippines where the carrier state and elimination of El Tor cholera vibrios lasted in convalescents up to several months or even a year. The vibrio identified in the bile ducts and gallbladder accounted for this situation.

This was first discovered in Egypt, in 1947, when in three of 30 convalescents studied vibrios were found in the bile collected by duodenal intubation; no vibrios were found in the stools of these three patients.

Until 1960 the cholera vibrio was isolated from the spontaneous stools or by rectal swab; a positive result was considered equal to the presence of the germ in the intestinal lumen. Since then, however, samples are more frequently collected after the administration of a purge, which considerably increases the chances of isolation since it causes a discharge of bile. The vibrio can thus be isolated even after longer intervals without a stool.

The presence of the vibrio in the bile had already been made evident in 1915, when isolated from the bile — after cholecystectomy — cholera vibrios in pure culture from a former cholera patient with negative stools.

Persistence of the cholera vibrios in the gallbladder and bile ducts as well as intermittent elimination of the vibrios in the faeces complicates the epidemiology of cholera very much, since it increases the duration and potential sources of infection and renders the detection of excretors more difficult.

The role of carriers in maintaining the epidemic or starting it again is more important than was at first believed, fully justifying the effort made at detecting them.

Today it has been accepted that whenever the import of the bacilli could not be established, long standing chronic carriers are responsible for the reappearance of the disease in areas in which it seemed to be extinct.

Another source, besides that of convalescent carriers, are the healthy carriers who have shown no clinical sign of disease. For each clinical case that appears in a community there are 10 to 100 asymptomatic cases. In El Tor epidemics the ratio is still greater. In a recent study in Pakistan the ratio of infection to disease was of 36:1 for the El Tor vibrio and 4:1 for the classical Inaba vibrio.

Healthy carriers have frequently been identified among the direct contacts of cholera cases and are designated in the speciality literature as *contact-carriers.*

The proportion of healthy carriers among the contacts has been studied by numerous authors in various epidemiological situations.

In Hong Kong a study included 108 contacts – the employees of a restaurant where one of the patients was working; 34 of these contacts excreted cholera vibrios without any clinical signs: 12 were detected after the first coproculture, nine after the second, seven after the third, three after the fourth, two after the fifth and one after the seventh coproculture.

This emphasises the fact that a cholera focus must not be considered to be extinct before multiple, long term investigations have been carried out.

Likewise in Hong Kong, in an El Tor cholera epidemic, 70 healthy carriers were detected among 1781 contacts. In another epidemic in Hong Kong the proportion of carriers was 0.34% to 9.51%; in Chiayi the proportion of contact-carriers (El Tor cholera) was 4.81%, and in the Philippines 18%.

The Centre for Cholera Research in Calcutta considers that 1.5% of the population are healthy cholera vibrio carriers.

The number of carriers among the contacts depends upon the intimacy of the contact. A study carried out in the Philippines showed that the proportion of carriers may reach 21% among the family contacts, 8.4% among the neighbours and 0.34% in the rest of the population in the locality.

In Bangladesh the proportion of healthy carriers among the members of the family hospitalised cholera patients was 30%.

The duration of vibrio excretion in the healthy contact carriers was on the average 2 to 14 days from the last contact. According to some authors it may last 30 to 40 days. Recent findings have shown that the duration of the contact carrier status may be within broad limits of a few months to 3 years. The case of a carrier infected more than 7 years previously has been reported.

The administration of antibiotics shortens very much the duration of excretion in contact carriers. In about 75% of the contact carriers treated the antibiotics curtailed the duration of excretion to less than 3 days.

Soviet authors have shown, however, that in the recent El Tor epidemic the treatment with antibiotics did not interrupt the carrier status. This appears to be characteristic of the El Tor biotype.

The difference between the classical and El Tor cholera also appears in the clinic. In spite of the known data concerning the frequency of severe forms, in the last pandemic there were few patients with the typical form of the disease, most had mild forms, in some instances a simple diarrhoea that lasted only a very short time. In the present El Tor pandemic the number of severe clinical forms was eight times less than in the epidemics with the classical vibrio.

Experience has proved that the mild atypical forms, very frequent at the beginning of an epidemic, represent the most active source of dissemination of the infection, because the patients consider they do not need any medical care and circulate among healthy people.

In view of the importance of carriers and patients the WHO together with the Indian Council for Medical Research in Calcutta conducted investigations with a view to detecting carriers and clinical cases of cholera by examining samples gathered from latrines. Each time a positive sample was found the stools of those who used the latrine were examined.

No excretors were found among the 632 persons using latrines in which the cholera vibrio was not detected; on the other hand, 54 carriers were found among the 433 persons using 200 positive latrines. Moreover, the study showed that a large number of persons eliminated El Tor vibrio even when there are few cholera patients. In endemic zones even when there are no clinical cases many people are infected, harbour and eliminate the germs for a long time, without presenting any of the clinical signs of the disease. This "iceberg" phenomenon is characteristic of intestinal infections and diarrhoeic diseases.

9. 2. Routes of transmission

The cholera vibrio penetrates into the human body by the digestive route by means of dirty hands (especially contaminated by faeces; vomit is of lesser importance), by drinking water (beverages, juices, etc.), food, crockery, cutlery, linen contaminated by patients or carriers, by flies, etc.

According to Gracias (1967) (Poona) the disease is more often transmitted by contaminated water than by food. In the endemic regions and those with sporadic cholera cases water plays a less important role than in epidemics; the transmission by dirty hands and contaminated food is more frequent in sporadical cases and an epidemic environment.

Epidemiologically, there are two kinds of epidemics:

(i) explosive epidemic with a large number of cases appearing within a short interval (1 to 15 days), with a common route of transmission (hydric as a rule);

(ii) a trailing epidemic in which a small number of cases is recorded daily or weekly. This route of transmission has not yet been clearly defined; it may be due to propagation from man to man.

In many regions the main factor is the lack of drinking water.

In endemic regions cholera presents characteristic seasonal variations. In Dacca, cholera appears after the rains of the monsoon have passed and disappears in the hot dry season; in contrast, in Calcutta the disease reaches a peak in the hot, dry season and decreases when the monsoon comes. These traits are characteristic of classical cholera.

The El Tor biotype differs from the classical vibrio not only in its clinical but also in its epidemiological expression.

A particular trait of the El Tor biotype which may influence the future of cholera epidemics is its great vitality and resistance outside the human body, particularly in foodstuffs.

In a well-documented work Felsenfeld (1966) studied the survival of cholera vibrios in foodstuffs, drinks, cutlery, crockery and other objects and showed that survival of the vibrios is, in general, longer than expected.

This suggests that indirect transmission of El Tor epidemics is favoured by resistance of the vibrio which is greater in nature and in food residues.

Transmission by water. Numerous examples are known of hydric transmission of cholera: the 1854 epidemic of London and the 1892 epidemic of Hamburg are classical examples; transmission by water was also incriminated in the cholera epidemic that broke out in Afghanistan in 1865.

Up to the use of chlorination, the epidemics in Calcutta were caused mainly by water transmission of the infection, particularly as surface waters were consumed.

Hydric transmission is linked to the relatively long survival of the vibrio in water, assessed to be 5 to 14 days in intensely polluted waters (residual waters) and somewhat more (according to some authors up to 21 days) in waters exposed to the sun. The survival time of vibrios in water, under natural conditions, is fairly relative since the presence of the bacteria may be due to continued pollution of the waters by contaminated faeces.

Survival of the vibrios in the water depends upon the conjugate action of several factors such as temperature, pH, salt content, the amount of organic matter in suspension and the degree of bacterial contamination.

Transmission by foodstuffs. A large variety of foods and drinks have been incriminated in the transmission of cholera.

Survival of vibrio cholera in food and beverages depends upon many factors such as temperature, humidity, pH, salinity, sugar content, some of which are favourable, others unfavourable.

Victor Babeş (1895) was the first to carry out systematic investigations on survival of the cholera vibrio in foodstuffs. He showed that the vibrios do not live more than 24 hours on fruit and vegetables with an increased acidity. On the other hand, on fresh fruit and vegetables with a low acidity survival of the vibrios may be prolonged 3 to 4 days (potatoes, spinach, green salad, raw cucumbers, onion, garlic, melons, dates, bananas). Meat and meat products may be a favourable substrate for cholera vibrios. Fish is likewise a comparatively favourable substrate: the vibrios resist 2 days in fresh fish, 1 day in smoked fish and at least 3 to 4 days in contaminated caviar kept on ice. In general, storage in the cold prolongs the survival time of the vibrios (up to 25 days in fresh-water fish stored at +3 to + 8°C). This is due to the remarkable resistance of the vibrios to low temperatures even below freezing point (it has been proved experimentally that the cholera vibrio survives more than 4 months in winter in the water of rivers).

In raw milk the cholera vibrio may survive 1—2 days, and much less in acidulated milk. Boiling destroys the vibrios in 5 minutes. In butter the survival is 14 days, but much less when the butter is slightly acid.

In cheese the vibrios maintain their vitality for 2 to 3 days. An increased salt content is not favourable to the cholera vibrio provided the pH is acid; on the other hand, a high saline concentration and alkaline pH represent an excellent condition for the survival of cholera vibrios.

In contaminated sugar the vibrios maintain their vitality for 2—3 days.

In coffee, tea, juices, etc. survival is shorter or less than 40 hours, according to some authors less than 24 hours; in beer and wine the vibrio lives only 3 hours and in pure lemon juice (pH 2.8) only 5 minutes.

These figures must, obviously, not be considered as absolute but only orientative.

It should be recalled that the viability of cholera vibrios in the environment — in water and food — is favoured by the cold, moisture and alkalinity, and lowered by acidity, heat, sunlight dryness and microbial competition.

Transmission through various objects. The currently contaminated objects by cholera vibrios only represent a danger when repeatedly contaminated. However, they may represent the role of a vehicle for short periods and within a restricted perimeter.

A particular danger for the transmission of cholera are the clothes and linen, bedlinen, carpets, blankets, mattresses, pillows, etc. soiled by excreta, especially when kept packed up in moist conditions. when the vibrios may survive up to 3 to 4 weeks.

Transmission through flies. As in all infectious diseases flies represent a means of transmission. They may transport the cholera vibrio from the excreta to food. The role of flies in the transmission of cholera has not been fully established but it does not seem to be very important.

Direct transmission has an outstanding role in the spread of cholera. From the experience gained in several epidemics it is considered that 30 to 40% of the cases occurred following direct contact with a patient or a carrier (Khan, 1967).

The appearance of a cholera case in an intact country or even in the neighbourhood of an area with cholera may give rise to a certain panic that often results in exaggerated measures of prevention known as "fear policy". The International Health Convention drawn up in 1851 by the great powers of that time tried to insure protection against this "disease of fear" in order to avoid the serious losses entailed by reducing the traffic of travellers and goods. The International Office of Public Health, the Sanitary Section of the League of Nations, and the WHO were created under the increasing pressure exercised by the medical problems and including cholera.

At present the specialists in cholera have drawn attention to the fact that anti-epidemic measures must keep account of the conditions in which the cholera vibrio survives in the environment. It is useless to take measures in excess of these considerations.

9. 3. Other factors that interfere in the epidemiology of cholera

The potentiality of cholera to spread over a certain area must not be determined according to the number of persons vaccinated but by the individual and collective state of hygiene and the efficiency of the sewage disposal and technicosanitary installations of the respective community.

The absence of healthy customs, under-nutrition, a high population density and high salinity of the water courses are factors that influence either separately or concomitantly the epidemiological mechanism of cholera.

The number of acute digestive diseases and especially of diarrhoeic syndromes that evolve in a community may create the "bed" in which cholera will run concomitantly or subsequently.

It is well known that the level of hygiene and public health of a given area is a decisive element that conditions whether onset of the first cholera case will or will not generate an epidemic and if this can be rapidly and definitely controlled or whether it will extend like an oil-spot.

Within an area in which the drinking water supply, alimentation and sewage disposal comply with the rules of hygiene, where the customs are healthy and there is a well developed medical network and all diarrhoeic syndromes carefully checked and treated the chances of an epidemic are considerably reduced. This was confirmed in Japan (1961—1963) and in England (1970). The imported cases, discovered in time did not trigger an epidemic owing to the thorough measures taken and to the high sanitary standard of these countries.

However, we must not overlook the fact that the El Tor pandemic differs from the preceding pandemics and may still hold a surprise. In a country in which diarrhoeic diseases are closely surveyed, a cordon sanitaire is instituted and the standard of sanitation satisfactory, there are many chances of arresting a cholera epidemic.

9. 4. Immunity in cholera

In infectious diseases clinical recovery is followed by a state of resistance to a renewed infection by the same bacteria. To this status there generally corresponds a certain specific antibody titre. In cholera, too, the increased resistance after the disease is accompanied by cholera antibodies in the blood. This state of immunity, weak and of short duration, may be explained by the poor contact between the body on the one hand, and the cholera vibrios and their toxins, on the other.

Immunity in cholera is not fully understood; most of the available data have been obtained experimentally in animals (dog, rabbit, etc.).

Our deficient knowledge of the immunological processes in cholera is due to our incomplete understanding of the pathogenic mechanism of the host-bacteria conflict taking place in the intestine and not going beyond it (Mosley, 1969).

Chronologically, it has been shown that after penetration of the vibrios in the small intestine (site of election) a multiplication of the pathogenic agent occurs. Consequent to multiplication of the vibrio in the intestine, vibriocide antibodies appear in the blood (of the IgM and IgG class), therefore a bacterial immunologic component. As regards the clinical expression of the disease cholera is not caused by bacilli but by the toxins (incompletely studied) produced under these circumstances. From the host-toxin conflict antitoxic antibodies (neutralising) belonging to the IgG class result, producing a toxic immunological component.

It is worthy of note that although the conflict takes place in the intestine — with an intact epithelium — it is reflected in the blood. What the relationship is between the two immunological components and the contribution of each one to the onset of immunity is not known. It is still more difficult to establish the correspondence between serum antibody titres and the state of immunity resulting from the local conflict. In cholera, however, antibodies also appear in the intestine, coproantibodies of the IgA class, especially secretory IgA, which is the dominant component. Today, the presence of secretory IgA has been proved in various secretions (colostrum, saliva, intestine, etc.), as well as the difference between it and IgA in the blood.

It is difficult to accept that serum antibodies, which are believed to be active only in the presence of certain complementary factors (lysosomes, etc.), may be active in the intestine where all the conditions are not met.

In cholera, it might be assumed that vascular permeability increases very much at the level of the intestinal mucosa, permitting passage of a sufficient number of elements necessary for the immunological reactions. Immunity in cholera appears to be better founded and explicable by local immunity (the existence of coproantibodies) rather than by serum immunity.

Another unsolved question is that of the relationship between the coproantibodies and the serum antibodies in this complicated mechanism of immunity in cholera. Investigations have been carried out in this connection and it was found that individuals with a sufficient vibriocide antibody titre contract the disease. The data obtained per age-groups in the endemic regions showed that infection may occur after birth and continues, the number of persons with vibriocide antibodies in the blood increasing by about 5% every year.

This natural stimulation of antibacterial immunity does not seem to depend only upon contact with the vibrio, since adults that leave the endemic regions for other cholera intact regions may become vibriocide antibody carriers in a small but significant proportion. These antibodies, qualified as natural, do not appear spontaneously, but seem to result following contact with related antigens *(Brucella, Pseudomonas)*.

The predisposing factors that determine that one person shall contract the disease and another pass only through a subclinical form of it are not sufficiently understood. Neither is the protective effect of passively transmitted antibodies well known, nor has it been established if a genetic basis exists in immunity to cholera.

Vaccination does not have a pathogenic ground because of our insufficient knowledge of the pathogenic mechanism and intimate immunological processes that govern this state of resistance. There is no available, efficient vaccine against cholera so far, which represents a serious drawback for prevention and control of the disease.

9. 5. Vaccination

Vaccines. The results obtained in more than 80 years of vaccination against cholera has not given a positive balance; both the efficiency and the duration of vaccination are unsatisfactory. The first attempt at vaccination (1885) is linked to the name of Feran (Barcelona) who used a vibrio culture in broth, injected subcutaneoulsy in small amounts. The live vaccine prepared by Feran did not induce the disease in laboratory animals and conferred resistance even to lethal cholera vibrio doses. Feran first tried the vaccine on himself, injecting 1 ml in the brachial triceps. He then vaccinated a great number of persons, irrespective of age. After 5 days he inoculated another 1.5 ml and then 2 ml in the same arm.

With this live vaccine more than 50 000 persons were vaccinated. Unfortunately, numerous Spanish and foreign specialists did not confirm the results reported by Feran. This explains why in the 1890 epidemic the Spanish authorities did not allow him to use the vaccine he had prepared.

The attempts of Haffkine in 1892 and 1913 to reconsider Feran's live vaccine yielded no results.

The killed vibrio vaccine was gradually introduced. However, the adopters of this method did not meet with significant success.

Remarkable were the results obtained by Cantacuzino (1920) during the cholera epidemic in 1913. The "Romanian experience" as this episode was called was not used in other epidemics although numerous trials were made at vaccination in cholera foci.

A cholera vaccination campaign was conducted in Madras in 1941–1943, with an inactivated vaccine but it did not considerably influence either the morbidity or the case fatality rate.

Since its foundation, the WHO has proposed to improve the specific prophylaxis of cholera. Between 1963 and 1966 six vast cholera vaccination campaigns were organised in Bangladesh, India and the Philippines, using 13 vaccines (administered in a single or double dose), all fluid except for that used in the Philippines in 1964 (vaccine in oil adjuvant) so as to be able to select a superior vaccine. In 1963, in Bangladesh, a highly endemic region, a significant protection of 18 months was obtained with a corpuscular vaccine. The following year, in full ascension of the epidemic, morbidity fell by 70% in the vaccinated community. Children under the age of 5 acquired lesser protection than adults. Similar results were obtained with a purified vaccine – a lipopolysaccharide of the Ogawa El Tor biotype.

It has been postulated that the dose should be augmented in children who are the most receptive to the disease in the regions of marked endemicity.

During the 1966–1967 vaccination campaign, likewise in Bangladesh, the efficiency of a classical vaccine (Ogawa strain N.I.H.$_4$ and Inaba N.I.H.$_{35}$ A_3 in equal parts) was tested in 40 000 children up to the age of 14, divided into three lots, one injection, two injections and a control lot with tetanus and diphtheria anatoxin.

Protection obtained with the vaccine was of 24% in the children under the age of 5 receiving one injection and 79% in those over the age of 5 receiving two injections. The duration protection did not exceed more than 3 months in either of the lots. The results were very similar to those of the Philippines in which 360 000 persons were vaccinated. Administered in a single dose but in a twofold concentration (16×10^9) the El Tor vaccine proved superior to the less concentrated vaccine administered in two doses ($2 \times 8 \times 10^9$) or one dose (at a concentration of 8×10^9). Although the twofold concentrated vaccine gave postvaccinal reactions certain specialists from the Kasauli Institute for Sera and Vaccines consider that higher vaccinal concentrations should be used.

In general, the results obtained with the fluid vaccines were very similar: a protection of 30 to 80% and a duration of 3 to 6 months; the reactogenic vaccines proved the most efficient.

The vaccine in oil adjuvant gave a long acting protection but severe local and general side effects and hence, was not recommended.

In the vaccination campaign of 1964–1965 in Calcutta an attempt was made to correlate the laboratory results (in mice) with field results. The corpuscular El Tor vaccine proved more active both in mouse and man than the classical vaccine (Philippines, 1964). The agreement between the protective effect in mouse and man was also noted with the vaccine used in Bangladesh.

For the preparation of the vaccine both the *V. cholerae* (a strain from the collection of the Bethesda National Institute of Health or local strains) and the El Tor biotype were used.

Improvement of the vaccine was also sought in other directions, a series of investigations being carried out on the role of antitoxic immunity. The experimental results obtained in dogs and rabbits are conclusive from this viewpoint.

The corpuscular vaccines used today cannot produce antitoxins, yet cholera with all its clinical attributes of gravity is caused by the vibrio toxins.

The opinion concerning the necessity of completing the bacterial antigenic component with a toxic component appears to be fully justified. This, however, implies a purified toxin isolated from the intestinal loop, which is almost impossible, especially as it is particularly labile and is rapidly degraded from the antigenic point of view, explaining why a cholera anatoxin has not yet been prepared.

Antitoxic immunity is also extremely difficult to attain since to be efficient the antitoxin must appear very soon in the intestine and in sufficient amounts to neutralise the toxin before it is fixed upon the sensitive tissues.

As the duration of immunity conferred by the present fluid vaccine is short the idea of using adjuvants to enhance and prolong this immunity arose. As already mentioned, the oily adjuvant (Arlacel) caused very severe reactions which were attributed to its subcutaneous instead of intramuscular administration. However, with Arlacel the accidents appeared to be due to a process of enzymatic hydrolysis.

In view of the superior effect of oily emulsions, especially metabolisable oils, other emulsifiers have been looked for. Aluminium compounds used in other vaccines appear to be indicated as, according to present findings, they enhance the immune reaction.

The attempts made at improving the present vaccine administered parenterally have not excluded the investigations carried out for establishing the conditions required for oral vaccination.

Oral vaccination with live attenuated vibrios, based upon the pathogenic mechanism of the disease (vaccination at the portal of entry) has been tried but special measures of safety have to be taken. To be accepted such a vaccine must fulfil several conditions: the strain must multiply in the intestine without producing the disease, it must be genetically stable from the apathogenic point of view, and it must possess several genetic markers permitting its precise identification.

The per oral vaccine prepared by Mukerjee with the W_6 streptomycin-resistant strain appears to satisfy these conditions. Moreover, the vaccine is perfectly tolerated.

In the intestine of the vaccinates (after one week) and in the blood (after 8 to 10 days) coproantibodies and vibriocide and antitoxic antibodies respectively appear. The strains isolated from the vaccinates present a good antigenic stability and unmodified virulence.

Investigations are continuing for the selection of an apathogenic strain (El Tor) and even avirulent mutants for either oral or parenteral administration, as well for the preparation of vaccines with antigenic fractions without the integral use of the vibrio.

A WHO report communicated that two purified protein antigens have been isolated which, injected separately, produce antitoxic antibodies and vibriocide antibodies respectively. Associated, these antigens might produce both kinds of antibodies.

On summing up the present data it may be stated that the immunological mechanism of cholera is still far from clear and the unsatisfactory results of vaccination depend upon numerous factors whose importance cannot be assessed.

Reliance must unquestionably be place on the pathogenic mechanism proper of the disease (differing in certain regards from other toxic infectious diseases) and for this account has to be taken of two facts of outstanding importance:

(i) the conflict takes place in the intestine (under conditions of an intact epithelium), a barrier that the cholera vibrio cannot pass and, therefore, does not reach the blood;

(ii) the disease is caused by the toxins, that cannot be overlooked in the pathogenic mechanism of the disease and especially the immune mechanism.

The local component, modified in case of repeated infections, produces a different reactivity in case of disease and in case of vaccination. On these grounds vaccination must be conceived as a complete method (both routes) and the vaccine must contain both antigenic groups (bacterial and toxic).

As local vaccination may achieve resistance at the portal of entry it should not be neglected irrespective of whether inactivated, live attenuated strains or antigens are used. Neither should the parenteral vaccine be dispensed with in connection with the study of a live attenuated vaccine.

A complete vaccination presupposes an approach by both routes and takes into account the two components (bacterial and toxic) and the immunological situation of the individual or the community as regards contact with cholera vibrios or antigenically related enterobacteriaceae; the immunological reactions are closely connected with this background against which the cholera vibrio antigens act.

Mode of administration. The fluid vaccines inactivated by heat or an antiseptic, recommended by the WHO, contain 8 milliard bacteria

The initial vaccination should be done in two doses, the second dose being injected 7 to 14 days after the first. For children the doses are reduced in terms of age; when strong reactions occur the doses are fractioned still further.

Vaccination must be repeated after 4 to 6 months in order to prolong the protection. The opinion of some specialists that the number of asymptomatic infections increase following vaccination is not based upon sufficient proof (The Committee of Experts, WHO, 1967).

●

The immunisation doses depend upon the age of the vaccinates:

TABLE 9. 1.

Age	Primary vaccination		Vaccination after 3—6 months
	1st dose	2nd dose	
1–2 years	0.2 ml	0.4 ml	0.4 ml
2–10 years	0.3 ml	0.6 ml	0.6 ml
over 10 years	0.5 ml	1 ml	1 ml

The interval between the first and second dose is of 7 to 14 days.

Contra-indications are:

(i) infectious diseases, in the acute stage and convalescence;
(ii) chronic cachectic diseases;
(iii) pregnancy;
(iv) severe system diseases;
(v) arterial hypertension;
(vi) allergic diseases of any kind.

The postvaccinal reactions may be local (pain, redness, infiltration pruritus) and/or general (fever, chills, stiffness) and disappear within 24 to 48 hours.

Accentuated local reactions, collapse and allergic reactions have been reported, and seem to be more frequent when large groups are vaccinated.

In cholera control today vaccination is considered an adjuvant method.

In the last El Tor pandemic vaccination did not help to limit the disease within the frontiers of the affected areas and was practically ineffective in dealing with the carrier state. Moreover, the vaccination certificate may be given by complaisance and the U.S.A. dismissed as unrealistic the expectation that travellers should have valid certificates of cholera vaccination. However, although vaccination is awarded lesser importance it is nonetheless compulsory according to the international health rules. The WHO considers that even if vaccination does not prevent the propagation of cholera it confers a certain degree of protection to those who enter endemic cholera regions.

9. 6. Measures of prevention and control

9. 6. 1. International provisions for cholera control

The pandemic character of El Tor cholera fully justifies an international programme for the control of cholera. The organ that coordinates and directs this activity is the WHO.

Such a programme should be determined by a consideration of the present particular epidemiological, clinical and therapeutical aspects of El Tor cholera. The WHO considers it necessary to take account of the following aspects:

(i) vaccination only confers partial immunity of short duration;

(ii) the detection and treatment of carriers is impossible;

(iii) survival of the cholera vibrio in foodstuffs is short, excepting milk and milk products;

(iv) The countries with an efficient sanitation system and a high level of hygiene may be considered non-receptive to cholera.

Operative, correct information and honest reporting must furnish the data which are rapidly processed, making it possible to recommend efficient measures. A certain reticence is observed in the notification of cholera cases, owing to the economic issues that may result.

The International Health Regulations apply the least possible restrictions on the international movements of travellers and goods. This implies an exchange of information between different states and especially neighbouring countries which ought to apply in common to measures aiming at cholera control.

Panic or exaggerated measures taken in some countries as well as non-compliance with the international rules of notification to the WHO complicate matters uselessly and make the situation more dangerous. The most adequate measures start with an attentive and constant epidemiological survey of:

(i) the endemic foci;

(ii) the most exposed neighbouring zones;
(iii) the susceptible zones with a variable degree of receptivity.

Epidemiological survey of the endemic foci and neighbouring areas. All countries are interested in, and were called upon to participate in the success of this survey. The main task falls to the lot of the respective countries that must promote a health organisation of their own, with hospitals, laboratories and epidemiological teams.

Only a complete system of recording and notification can lead to a thorough knowledge of the situation, efficient measures and provisions.

In some of the affected countries the working conditions are very difficult and the financial aspect burdensome. However, the efficiency of the epidemiological survey will increase if combined with that of other acute gastrointestinal diseases, especially salmonelloses and diarrhoeic infections. All asymptomatic cases ought perhaps to be admitted to quarantine but for this more sensitive bacteriological tests need to be found, more financial support and a change in the attitude of the population which is not very helpful in this connection.

In spite of these difficulties epidemiological survey, in the endemic zones must constantly keep in view the detection and notification of the cholera cases, informing the neighbouring countries and the WHO.

Epidemiological survey in susceptible areas (countries) with a variable degree of receptivity. These countries must organise a rapid detection programme, keeping the neighbouring countries informed concerning the evolution of eventual cholera cases and if necessary establishing control frontier points and along the lines of communication in order to prevent import of the disease, without taking excessive or useless measures of prevention, the programme should have in view:

(i) the survey of all diarrhoeic diseases, especially at ports, airports and frontier stations, as well as along the lines of communication within the country;

(ii) medical control units for the isolation and treatment of suspects at the frontier stations.

(iii) organisation of a network of laboratories;

(iv) medical control of travellers and goods from infected areas.

9. 6. 2. Cholera prevention and control in intact countries

The last El Tor pandemic broadened the perimeter of vulnerable countries. Today all countries irrespective of their geographical situation and health organisation may be susceptible to the import of cholera. In order to prevent this and to control the disease in as short an interval as possible each country must draw up a plan of action providing for every contingency. The more operative and competent the smaller is the risk of import of the disease and its spread in the country.

The national plan for the control of cholera includes three distinct stages corresponding to the measures that have to be taken:

(i) the preparatory stage with:

(a) permanent measures, and

(b) special measures in case of immediate threat;

(ii) invasion (epidemic) stage;

(iii) postepidemic stage.

To each stage there corresponds complex technical measures and organisation.

9. 7. Preparatory stage

The fact that, starting in 1961, cholera manifested a tendency to a pandemic evolution justifies the preoccupation of each country in bringing up to date its methods of cholera control and to draw up preparatory measures.

In the preparatory stage it is of the utmost importance that health officials should not underrate the possible import of cholera. In this stage, every country must survey the frontier points and at the same time be prepared technically and with the necessary material to be able to start at any moment action against cholera.

The plan of action must be drawn up in keeping with the provisions of the international Health Regulations referring to quarantine diseases (El Tor cholera included).

For persons coming from cholera regions the following rules should also be complied with:

(i) Foreign citizens arriving in the country from infected regions must show their international vaccination certificate (within the previous 6 months). If less than 7 days have passed since they left the contaminated area they are given tetracyclin preventively (2 g/24 h for 3 days) and are kept under medical surveillance for 5 to 7 days.

(ii) Citizens travelling in contaminated areas are vaccinated before starting (their vaccination certificates are checked at the frontier points before starting).

(iii) Citizens coming from cholera regions and not vaccinated are kept in quarantine for 5 days. Irrespective of whether they were vaccinated or not they will be checked by coproculture after taking a purge (magnesium sulphate — 15 g) and treated with tetracyclin (2 g/24 h for 3 days). They will be kept under medical surveillance and controlled periodically. Persons with acute digestive disturbances are admitted to hospital, controlled bacteriologically and treated with tetracycline (2 g) for 3 to 5 days.

(i) Vaccination of the personnel working at the frontier points (customs officials, frontier guards, etc.).

(ii) Instruction of the medical personnel working at frontier points, ports or airports, tourist centres and communicable diseases hospitals, concerning;

(iii) the particular clinical aspects of cholera (mild, atypical, subclinical forms, etc.);

(iv) emergency treatment of cholera patients.

(v) The training of a special team of doctors and ancillary staff which will be vaccinated and ready at any moment to apply the treatment in case of imported cholera cases.

In the choice of this personnel the criteria of competence, professional trustworthiness, energy and calm must be taken into consideration in the first place.

(i) The hospitals that will serve for the cholera cases must be equipped in due time with the necessary materials and facilities.

(ii) The hospital laboratory must be able to perform at any moment a bacteriological diagnosis which will be confirmed by the central reference laboratory.

(iii) The sanitary organisation and especially the anti-epidemic network must be instructed so as to be able to take action if a cholera case appears.

(iv) Central organisation, stockpiling and renewal of perfusion fluid, vaccines, antibiotics, etc. to be sent to any locality when necessary.

(v) Information concerning the development of cholera throughout the world and constant contact with the neighbouring countries concerning the measures to be taken.

In case of an imminent threat of contagion special sanitary anti-epidemic measures for the prevention of cholera are instituted.

The instructions of the sanitary authority should foresee:

(i) Health control on arrival of all travellers coming from contaminated countries.

(ii) Permission to enter the country of persons coming from, or that have passed through, contaminated countries only if they have an international vaccination certificate.

(iii) Isolation in special units at the frontier points of suspects up to their evacuation to hospital.

(iv) Persons possessing a cholera vaccination certificate and coming from contaminated countries must give the address at which they will be staying for the next five days and the officials at the frontier must notify the militia and health inspectorate of the respective district. Travellers who cannot say where they will be staying must accept quarantine in the nearest communicable diseases hospital for three days during which they will be dosed with tetracyclin (2 g per day).

(v) Strangers without relatives in the country receive accommodation in hotels. The hotel personnel must give the district health inspectorate the names of travellers coming from contaminated countries.

(vi) The access of foreign sailors coming from contaminated areas to the sea and river ports will be restricted even if they have a vaccination certificate, that is:

(a) landing is allowed only to persons who are indispensable for formalities and commercial transactions;

(b) the crew can land only after bacteriological control of the stools (coproculture);

(c) restriction to a minimum of the number of persons strictly needed for loading and unloading;

(d) the captains of seagoing vessels will be instructed concerning the means of avoiding infection; on coming home they can only land after bacteriological control.

(e) periodical disinfection of the vessels plying to and from contaminated countries.

(vii) Interruption of tourist excursions to contaminated areas.

(viii) Intensification of the measures of individual and collective hygiene by:

— a good, controlled supply of drinking water;

— removal of residues and their processing;

— sanitation of the market places, common transport means and sites of agglomeration;

— sanitation and maintenance of the hygiene of units for the production and sale of foodstuffs;

— strict compliance with individual hygiene of the persons producing, handling or selling food, and equipment for the respective units with the necessary material for maintaining proper health conditions and disinfection;

— the employees of the alimentary sector and those working with the water supply system must notify the health officials when suffering from acute digestive disturbances; they will not be allowed to work.

Fruit and not heat-processed food imported from cholera areas will be sold only 5 days after being sent off from the respective country. Fruit, vegetables, beverages and food not prepacked found at the customs on travellers from contaminated areas will be destroyed.

All sanitary units irrespective of their administrative subordination can be used for cholera vaccination of the groups indicated by the Ministry of Health and for application of the technical and organisational measures established in the instructions.

Moreover, in view of the mild, atypical forms, the measures taken for the detection of cholera have to be broadened including:

Bacteriological control (for the cholera vibrio) of all acute digestive infections and especially diarrhoeic diseases. The hospital laboratories are obliged to perform a routine diagnosis for cholera, Shigella and Salmonella infections in all cases of diarrhoeic diseases.

Testing large groups of the population in the districts exposed to cholera contamination by mass coprocultures.

Close relationships with the neighbouring countries with a view to taking common measures of prevention and control.

9. 8. The invasive stage of cholera

The measures taken in this stage are based upon the general principles of anti-epidemic control against enteric diseases.

When a cholera case appears the measures to be taken keep in view the patients, the carriers and the contacts, as well as a series of general measures for restricting the spread of cholera.

9. 8. 1. Measures concerning the patients

As man is the only source of transmission of cholera in a community, early rapid detection and isolation of the patients plays a primary role in arresting an epidemic.

The fact that a cholera patient, irrespective of the clinical form, excretes 10^8—10^9 and a carrier 10^2—10^5 cholera vibrios/ml in the faeces convincingly illustrates the high propagation ability of the infection.

The detection of cholera patients must take into account certain essential elements in the mechanism spreading the infection, namely:

(i) the short period of incubation, permitting rapid multiplication of the cases within the infected area;

(ii) the high incidence of mild forms (common enterocolitis) especially in children making the detection of atypical cases very difficult.

At present, now that El Tor cholera has replaced classical cholera the possible spread and consolidation of the disease has increased considerably. In view of these facts a mass action for the detection of patients and carriers must be started by means of:

(i) operative epidemiologic survey, delimiting the probable area of extent of the disease;

(ii) daily checking (or even twice a day) from house to house (the organised communities included) for the identification of patients;

(iii) Health education by the press, broadcasting stations and TV, showing the importance of coming to medical units at the first signs of the disease;

(iv) medical examination, including laboratory tests of anybody with clinical signs of an enteric disease.

Any clinically suspect individual is admitted to the cholera unit and isolated. A patient can only be discharged after clinical and bacteriological control (established after three coprocultures performed at 3 to 4 days interval or, as in India, during the last three days before discharge).

Apart from the patients, carriers must also be detected and treated.

During an epidemic a greater number of patients may render a selection necessary: those with severe forms will be admitted to hospital and the others treated at home by mobile medical teams.

9. 8. 2. Measures concerning the carriers

Cholera vibrio carriers will be admitted to hospital until they have completely recovered bacteriologically, proven by three negative coprocultures.

During epidemics, when this is not possible, they will be detected and treated at home.

For sterilisation of cholera vibrio carriers tetracyclin and numerous other antibiotics and chemotherapeutical drugs are used (Mc Cormack *et al.,* 1968); streptomycin administered per oral determined sterilisation of 96% of the carriers. Sulphonamides, long acting with slow elimination, have likewise been tested.

In general, sterilisation of carriers is more difficult than antibiotic treatment of the patients.

The role of carriers in the spread of the disease and the importance of prophylaxis in communities with a high risk of infection (persons coming in contact with carriers or patients) must be emphasised once again (Barna *et al.,* 1967).

9. 8. 3. Measures concerning contacts

The contacts established by epidemiological survey will be isolated and surveyed clinically and bacteriologically (coprocultures) during 10 days after their last contact with the patient.

Chemoprophylaxis is indicated in all family contacts. This measure is efficient the sooner the patient is discovered as the incubation period is very short.

When cholera vibrios are isolated or clinical signs of the disease appear (even in the absence of vibrios in the stools) the contacts are considered cholera suspects or carriers and treated as such,

For the direct contacts (family, home, etc.) an antibiotic treatment is recommended (2 g tetracyclin 3 days running); a 24 hour treatment does not give results.

For indirect contacts (living on the same street, village, etc.) cholera vaccination or revaccination is recommended.

9. 8. 4. Measures concerning the routes of transmission

In a cholera epidemic the drinking water supply, food circuit, sewage disposal, individual and collective hygiene must be closely controlled by the health authorities aided by the administrative organs and the population.

In localities with a central water supply system, bacteriological control of water and chlorination must be done daily. In villages where the water is drawn from wells, the water for drinking and household needs must be boiled.

Throughout the epidemic raw milk and derivatives are forbidden, as well as raw fruit and vegetables, ice-cream and natural ice. In general, it is recommended that only heat-processed food should be consumed.

During the epidemic the personnel working at the central water supply installations will be subject to clinical, epidemiological and bacteriological control, as well as those working in the food sector for early detection and isolation of eventual patients, carriers or contacts.

Special attention must be paid to the dumping and transport of household refuse; the latrines must be constantly disinfected. A repeated systematic action for the destruction of flies must be initiated.

During cholera epidemics the measures for neutralisation of the routes of transmission of the disease stem from the basic principles for the control of enteric communicable diseases (typhoid fever, dysentery, etc.), allowing however the peculiarities of the cholera vibrios (viability in water, food, various pH, temperatures, degree of moisture, etc.).

9. 8. 5. Measures concerning the receptive population

When there is a cholera epidemic outbreak the entire population must be vaccinated. The movement of the population from the infected areas to intact regions and the reverse will be restricted as much as possible.

A debatable question is that of prevention with antibiotics. This has been applied in the course of certain epidemics, dosing the population with chloramphenicol, for instance in Iran and Japan.

The WHO considers that the systematic administration of antibiotics to the mass of the population in order to restrict spread of the infection involves severe risks and should not be used.

Antibiotics may change the intestinal flora of healthy subjects rendering them still more prone to contract the infection. Nor should the danger of antibiotic resistant strains be minimised. Account must also be taken of possible accidents following mass chemoprophylaxis with long-acting sulphonamides of the Fanasyl type (although readily administered once a day), as illustrated by the experience in Africa.

Prophylaxis with antibiotics is indicated therefore for family contacts, i.e. where a high risk is involved.

9. 9. The postepidemic stage

A strict clinical and bacteriological survey of enteric diseases must be continued in order to detect the presence of mild, atypical forms. When cholera cases or carriers fail to be detected cholera may become endemic in the respective area.

In Uzbekistan a vast epidemiologic survey of the endemic zone disclosed wide circulation of the cholera vibrio among the population many months after the epidemic had been arrested. This points to the necessity of continuing the survey of an area once affected by cholera.

9. 10. Disinfection

Among the measures of prevention and the control of cholera, disinfection occupies a very important place. Disinfection has a wide field of applicability and must lead to inactivation of the source of infection and interruption of the routes of transmission.

A first emergency is disinfection of the patient's stools and vomit, the faeces of carriers and contacts in the focus.

To this may be added: disinfection of:

(i) the patient's wearing apparel, linen and bedclothes (sheets, pillowslips, pillows, blankets, mattresses, covers, etc.);

(ii) the patient's bed, the walls of the room and floor;

(iii) the dishes, cutlery, handkerchiefs, toothbrush, bedpan, etc.;

(iv) food residues, etc.;

(v) the objects used by patients, carriers or contacts;

(vi) the medical instrumentation (catheters, tubes, thermometers, kidney dishes, drums, etc.).

The cholera vibrios are destroyed by:

(i) burning of objects of no value (newspapers, books, toys, cloths, etc.);

(ii) boiling for 30 minutes (linen, clothes, crockery, etc.);

(iii) chemical disinfectants of the chlorigen class, cetyl pyrimidium bromide (Cetazolin, Bromocet), etc.

In order to obtain good results disinfection must be well organised, in successive stages and founded on a basic understanding of the proceedings.

Correct application of these measures involves no random choice of the methods to be used, but, on the contrary, a thorough knowledge of the biological properties of the causal agent and the characteristics of the disinfected object.

The resistance of the cholera vibrio to environmental factors is not very great (that of the El Tor vibrio somewhat greater) but sufficient to be considered an epidemiogenic threat (Miyaki *et al.*, 1967).

The cholera vibrio resists the cold well: in ice it lives for several months and in water 5 to 14 days, in terms of pH, temperature, the amount of organic matter, microbial flora, etc. In residual waters it resists 6 to 7 days and in the water of sources 2 to 3 weeks.

In foods the vibrio may last 10 to 12 days in terms of the same parameters.

In the presence of fermentation and putrifying germs it is rapidly destroyed. On soiled linen the vibrio may resist 12 days and on dry linen 2 days. In sunlight it lasts about 2 hours.

The cholera vibrio is sensitive to heat: at a temperature of 50 °C it is killed in 60 minutes, at 100 °C instantaneously; pasteurisation at 50 to 60°C kills the vibrio in 20 to 30 minutes, formaldehyde (5%) and phenol (1%) in 1 minute.

As the germs are to be found in pathological products rich in organic matter, a good disinfection implies a longer contact between the germs and the disinfectant.

The following methods of disinfection are used:

Disinfection of the excreta (excreta, diarrhoeic, stools, urine, vomit), collected in a container with a lid, will be mixed with dry chlorinated lime in a proportion of 1 : 5 of the weight of the excreta (200 g per litre), or 2% lime milk (twice the amount of the excreta) or free chlorine. The contact time in closed containers is 2 hours. The bedpans or chamber pots are introduced each time after use in 0.5% chlorinated lime or 0.5% chloramine solution, for 30 minutes.

Disinfection of the hands is done with a 1 : 5000 potassium permanganate solution (1% Bromocet), detergent (2.5%) and ethylic alcohol (25%) or an iodine water alcohol solution of iodine (2 g), potassium iodide (2.4 g), propylenglycol (50 g) and distilled water (up to 100 ml). After chemical disinfection the hands are washed with soap and water, rinsed and dried with hot air or a disposable paper towel.

Disinfection of the linen and bedclothes. The linen is collected in the patient's room, put in a container with a lid, to which a disinfectant is added (preferably 0.2% chloramine). The container is then carried to the disinfection unit of the annexe and from there to the laundry. After one hour's disinfection the linen is boiled for 40 minutes, adding 1% soda to the water. After washing and ironing the linen is introduced in individual, well-closed polyethelene bags.

The mattresses, pillows, blankets, dressing gowns, etc. are autoclaved for 6 to 8 hours (steam with 20 cm^3 40% formaldehyde per 1 m^3).

Disinfection of the bed is done daily with a 2% bromocet solution at the same time as disinfection of the floor and walls.

Disinfection of the slippers. The slippers are kept for 15 minutes in 1% bromocet solution, then the soles wiped with 0.5% chloramine.

The disinfection of surfaces. Oil painted floors are washed with 0.5% chloramine solution, using about 250 cm^3 per 1 m^2 so that the floor should remain wet for 15 minutes. Boards must be scrubbed with the brush with 200 cm^3 1% chloramine and left to dry for 20 to 30 minutes.

The walls are whitewashed twice with 20% lime milk; this is repeated every 8 days in the wards and admission service and every 18 to 21 days in the rest of the hospital.

Disinfection of the water closets. The water closets are wiped with a 10% chlorinated lime or 0.5% chloramine solution. It should be recalled that the rooms must be aired about 2 hours after using chlorinated lime.

Disinfection of the crockery and cutlery. The best method is to boil them with 1—2% soda for 30 to 40 minutes. If this is not possible then chloramine (5—6 g in bucket of water) or chlorinated lime (a spoonful and a half in a bucket) is used. The dishes must be left at least one hour in contact with the disinfectant. Metal objects and cutlery can only be disinfected by boiling. Irrespective of the methods or means used everything must be rinsed under a hot water jet.

The food remaining on the dishes must be emptied into a container with chlorinated lime 1 : 5 or lime milk 2 : 1 and kept for 1 to 2 hours before emptying them. The solid remains are burnt if possible. The vessels in which the food was brought to the patient are only taken back to the kitchen after disinfecting them with 10% lime milk.

Larder. The floor is washed daily with lime milk and every 3 to 4 days with a warm 2% soda water. The larder must well aired and whitewashed every 2 weeks with a 20% lime milk solution.

Disinfection of the laboratories. The floor, tables, cupboards are washed daily wiping them with a 2% bromocet solution every 2 to 3 days with 0.5% chloramine.

Disinfection of water. Tap water usually contains 0.05—0.5 mg residual chlorine per litre. For the disinfection of smaller amounts of water one tablet chloramine (0.5 chloramine, or 0.25 g active chloride) is added to 10 litres water.

During epidemics boiled water alone will be used.

Ultraviolet light is used as an adjuvant only after chemical disinfection. Irradiation is done for 20 minutes, as a rule in the laboratories and pharmacy.

The *cadavers* of cholera patients are taken to the mortuary of the cholera hospital (bungalow), protected against flies and handled only by vaccinated and carefully instructed personnel.

The cadavers are washed and dressed, covered in a cloth treated with a disinfectant sublimate 1%, cresol 5% or phenol 3—5%). It is recommended to put a layer of quick lime at the bottom of the coffin. The lid must be nailed on and sealed.

References

Babeş, V. (1885) *Virchows Arch. Path. Anat.*, *99*, 148.
Bart K. J., Kahn M., Mosley W. H. (1970) *WHO Bull.*, *43*, 421.
Barua D. (1970 a) *Public Health Papers WHO*, *40*, 29.
Barua D. (1970 b) *Public Health Papers WHO*, *40*, 47.
Barua D., Gomez C. Z. (1967) *WHO Bull.*, *37*, 798.
Barua D., Wake A., Gomez C. Z., Paquio A., Azurin J. C., Dizon J. J., Ramos R., Cordova V. (1967) *WHO Bull.*, *37*, 804.
Basu S., Mukerjee S. (1970) *WHO Bull.*, *43*, 509.
Benenson A. S., Islam M. R., Greenough W. B. (1964) *WHO Bull.*, *30*, 827.
Benenson A. S., Mosley W. H. (1968 a) *WHO Bull.*, *38*, 277.
Benenson A. S., Saad A., Paul M. (1968 b) *WHO Bull.*, *38*, 267.
Benenson A. S., Saad A., Mosley W. H., Ahmed A. (1968 c) *WHO Bull.*, *38*, 287.
Benenson A. S., Joseph P. R., Oseasohn R. O. (1968 d) *WHO Bull.*, *38*, 347.
Benenson A. S., Mosley W. H., Fahimuddin M., Oseasohn R. O. (1968 e) *WHO Bull.*, *38*, 359.
Burrows W. (1968) *Annual Review of Microbiology*, *22*, 245.
Cantacuzino J. (1920) *Ann. Inst. Pasteur*, *34*, 57.
Carpenter C.C.J. (1969) *Texas Reports on Biol. and Med.*, *27*, suppl., 203.
Carpenter C.C.J., Mondal A., Sack R. B., Mitra P. P., Dans P. E., Wells S. A., Hinman E. J., Chaudhuri R. N. (1966 a) *Bull. John Hopkins Hosp.*, *118*, 174.
Carpenter C.C.J., Barua D., Wallace C. K., Mitra P. P., Sack R. B., Kharra S. R., Wells S. A., Dans P. E., Chaudhuri R. N. (1966 b) *Bull. John Hopkins Hosp.*, *118*, 216.
Carpenter C.C.J., Sack R. B., Mitra P. P., Mondal A. (1964) *Bull. Calcutta School Trop. Med.*, *12*, 30.
Craig J. P. (1965) *Nature*, *1*, 207.
Cash R. A., Nalin D. R. (1970) *Amer. J. Trop. Med. Hyg.*, *19*, 653.
Chaudhuri R. N., Neogy K. N., Sanyal S. N., Grupta R. U., Manji P. (1968) *Lancet*, *1*, 332.
Das Gupta A., Sinha R., Shrivastava D. L., De S. P., Taneja B. L., Rao M. S., Abon-Gareb H. A. (1967) *WHO Bull.*, *37*, 371.
D'Hérelle F., Malone R. H., Lahri M. N. (1930), *Indian Med. Res. Mem.*, *14*.

Doorenbos W. (1932), *Ann. Inst. Pasteur*, *48*, 457.
Dutta N. K., Panse M. U. (1963) *WHO Bull.*, *28*, 357.
Dutta N. K., Panse M. U., Kulkarni D. R. (1959-1960) *Rev. Proc. Diamond Jubilee, Hofkine Inst.*, 44.
Dutta N. K., Oza N. B. (1963) *WHO Bull.*, *28*, 307.
El Ramli A.H., (1948) *J. Roy. Egypt. Med. Assoc.*, *31*, 322.
El-Shawi N., Thewaini, A. J. (1969) *WHO Bull.*, *40*, 163.
Feeley C. J., Pittman M. (1963) *WHO Bull.*, *28*, 347.
Feeley C. J., Robers C. P. (1969) *Texas Reports on Biol. and Med.*, *27*, suppl., 213.
Felix H. (1971) *Presse Méd.*, *79*, 475 and 1801.
Felsenfeld O. (1965) *WHO Bull.*, *33*, 725.
Felsenfeld O. (1966) *WHO Bull.*, *34*, 161.
Finkelstein R. A. (1969) *Texas Reports on Biol. and Med.*, *27*, suppl., 181.
Finkelstein R. A., Raungthum A. J. (1968) *WHO Bull.*, *38*, 803.
Forbes G. I., Lokhart J.I.F., Bowman R. K. (1967) *WHO Bull.*, *36*, 367.
Fuhrman G. J., Fuhrman F. A. (1960) *Nature*, *188*, 71.
Gallut J., Nicolle P. (1963) *WHO Bull.*, *28*, 389.
Gallut J., Quinon J. (1970) *WHO Bull.*, *42*, 464.
Gangarosa E. J., Beisel W. R., Bengajati C., Sprinz H., Piyaraty P. (1960) *Amer. J. Trop. Med. Hyg.*, *9*, 125.
Gangarosa E. J., Saghari H., Emile J., Siadat H., Watanabe J. (1966) *WHO Bull.*, *35*, 669.
Gangarosa E. J., Bennet J. V., Voring J. R. (1967) *WHO Bull.*, *36*, 897.
Gardner A. D., Venkatraman K. V. (1935) *J. Hyg.* (London), *35*, 262.
Ghosh A. K., Kasai G. J., Burrows W. (1969) *WHO Bull.*, *39*, 964.
Gotschlich F. (1906) *Z. Hyg. Infek. Kr.*, *53*, 281.
Gracias R. (1967) *An Introduction to the Diagnosis. Treatment and Control of Infectious Diseases*, St. Patrick's Press, Bishop's House, Poona, 1967.
Grady G. F., Chang M. C. (1970) *J. Inf. Diseases*, *121*, suppl., 92.
Greenough W. B., Gordon R. S. Jr., Rosenberg I. S., Davies B. I., Benenson A. S. (1964) *Lancet*, *1*, 355.
Heiberg B. (1934) *C. R. Soc. Biol.* (Paris) *115*, 984.
Hirshhorn H., Chaudhuri A.K.M.A., Lindenbaum J. (1969) *Lancet*, *1*, 1230.
Hugh R. (1965) *Int. Bull. Bact. Nomencl.*, *15*, 61.
Kenneth B. J., Moslemudin K., Mosley W. H. (1970) *WHO Bull.*, *43*, 421.
Khan A. Q. (1967) *Lancet*, *1*, 245.
Kobari K., Uylangco C., Vasco J., Takahira J., Shimizu N. (1967) *WHO Bull.*, *37*, 751.
Kobari K., Uylangco C., Vasco J. (1967) *WHO Bull.*, *37*, 810.
Kobari K., Takakura I., Nakatomi M., Sogame S., Uylangco C. (1970) *WHO Bull.*, *43*, 365.
Koh L., Tan T. H., Tijia S. K., Gan K. H. (1967) *WHO Bull.*, *36*, 342.
Kuwahara S., Goto S., Kimura M., Abe H. (1967) *WHO Bull.*, *37*, 763.
Lindenbaum J., Robert S. G., Hirshhorn H., Akbar K., Greenough W. B., Islam M. R. (1966) *Lancet*, *1*, 1066.
Lindenbaum J., Greenough W. B., Islam M. R. (1967 a) *WHO Bull.*, *37*, 529.
Lindenbaum J., Greenough W. B., Islam M. R. (1967 b) *WHO Bull.*, *36*, 871.
Mahalanabis D., Sack R. B., Kaplan J., Jacobs B., Mondal A. (1970) *WHO Bull.*, *42*, 837.
Mc Cormack W. M., Chaudhuri A. M., Jahangir N., Fariduadin A. D., Massey W. A. (1968) *WHO Bull.*, *38*, 787.

Miyaki K., Iwahara S., Sato K., Fujimoto S., Aibara K (1967) *WHO Bull.*, *37*, 773.

Monsur K. A., Rahman M. A., Huc F., Istam M. R., Northrup R. S., Hirshhorn H. (1970) *WHO Bull.*, *42*, 723.

Monsur K. A., Rizvi S.S.M., Huq M. I., Benenson A.S. (1965) *WHO Bull.*, *32*, 211.

Mosley W. H. (1969) *Texas Reports in Biol. and Med.*, *27*, 227.

Mosley W. H., Ahmad S., Benenson A.S., Ahmed A. (1968) *WHO Bull.*, suppl., *38*, 777.

Mosley W. H., Mc Cormack W. M., Fahimuddin M., Azis K.M.A., Rahman M.A.S.M., Chaudhuri A.K.M.A., Martin A.R., Feeley C. J., Phillips R. A. (1969) *WHO Bull.*, *40*, 177.

Mosley W. H., Mc Cormack W. M., Ahmed A., Chaudhuri A. K. M. A., Barui R. K. (1969) *WHO Bull.*, *40*, 127.

Mukerjee S. (1963 a) *WHO Bull.*, *28*, 333.

Mukerjee S. (1963 b) *WHO Bull.*, *28*, 337.

Mukerjee S. (1970) *Public Health Papers WHO*, *40*, 39.

Mukerjee S., Basu S., Bhettcharya P. (1965) *Amer. J. Trop. Med. Hyg.*, *14*, 780.

Nalin D. R., Cash R. A., Islam M. R., Molla M., Phillips R. A. (1968) *Lancet*, *17*, 370.

Nalin D. R., Cash R. A., (1970) *Cahier de Santé Publique no. 40 OMS*, 73.

Nalin D. R., Cash R. A., Rahman M. (1971) *WHO Bull.*, *43*, 361.

Napier L. E. (1951) *Cholera in Modern Practice in Infectious Fevers*, Banks H. S. ed., New York.

Neogy K. N., Mukherji A. C. (1970) *WHO Bull.*, *42*, 638.

Nestorescu N., Popovici M., Popescu Al., Georgescu C. (1965) *Rev. San. Mil.*, *61*, no. special, 305.

Oza N. B., Dutta N. K. (1970) *WHO Bull.*, *42*, 642.

Pandit C. G., Pal S. C., Murti G. V., Misra B. S., Murty D. K., Shrivastiav J. B. (1967) *WHO Bull.*, *37*, 681.

Pasricha C. L., De Monte A.J.H., Paul B. M. (1936) *Indian Med. Gaz.*, *71*, 61.

Pesigan T. P., Plantilla L., Rolda M. (1967 a) *WHO Bull.*, *37*, 779.

Pesigan T. P., Gomez C. Z., Barua D. (1967 b) *WHO Bull.*, *37*, 814.

Phillips R. A. (1963) *WHO Bull.*, *28*, 297.

Phillips R. A. (1967) *J. Amer. Med. Assoc.*, *202*, 610.

Pierce N. F., Banwell J. G., Mitra R. C., Caranasos G. J., Keimowitz R. T. (1970 c) *Annals of Internal. Med.*, *72*, 1959.

Pierce N. F., Banwell J. G., Mitra R. C., Caranasos G. J., Keimowitz R. T., Thomas J. C., Mondal A. (1968) *Brit. Med. J.*, *3*, 277.

Pierce N. F., Banwell J. G., Sack R. B., Mitra R. C., Mondal A. (1970 a) *J. Inf. Diseases*, *121*, suppl., 31.

Pierce N. F., Banwell J. G., Gorbach S. L., Mitra R. C., Mondal A. (1970 b) *Annals of Internal Med.*, *72*, 357.

Pollitzer R. (1959) *Cholera, WHO Monograph Series no. 43*, 58.

Popescu Al., Georgescu C. (1965) *Rev. San. Mil.*, *61*, no. special, 307.

Prescott L. M., Datta A., Datta G. C. (1968) *WHO Bull.*, *39*, 917.

Prescott L. M., Battachargee N. K. (1969) *WHO Bull.*, *40*, 980.

Richardson S. H., Moftle K. A. (1970) *J. Inf. Diseases*, *121*, suppl., 73.

Rizvi S., Benenson A. S. (1966) *WHO Bull.*, *35*, 675.

Rogers L. (1915) *Lancet*, *2*, 219.

Rogers L. (1952) *Cholera in Tropical Medicine*, ed. by Rogers L. and Megaw J.W.D., sixth edition, London.

Sack R. B., Cassels J., Mitra R., Merritt C., Butler T., Thomas J., Jacobs B., Chaudhuri A., Mondal A. (1970) *WHO Bull.*, *43*, 351.
Sanyal S. C., Mukerjee S. (1969) *WHO Bull.*, *40*, 503.
Sayamov R. M. (1963) *WHO Bull.*, *28*, 311.
Takeya K., Shimodori S., Zinnaka Y. (1967 a) *WHO Bull.*, *36*, 990.
Takeya K., Shimodori S., Gomez C. Z. (1967 b) *WHO Bull.*, *37*, 806.
Uylangco C., Vasco J. (1967 a) *WHO Bull.*, *37*, 810.
Uylangco C., Vasco J., Calilong V. (1967 b) *WHO Bull.*, *37*, 812.
Wallace C. K., Anderson N. P., Brown T. O., Khandra S. R., Lewis G. V., Pierce N. F., Sanyal S. C., Segre C. V., Waldman R. H. (1968) *WHO Bull.*, *39*, 239.
Watten R. H., Phillips R. A. (1960) *Lancet*, *2*, 999.
Zafari Y., Zarifi A. H., Rahmanzadeh S., Sanyan H. (1968) *WHO Bull.*, *39*, 492.

INTERNATIONAL HEALTH REGULATIONS

(1969, *amended* 1973)
World Health Organisation
Geneva, 1974

SMALLPOX

Art. 7 — 2. An infected area may be considered as free from infection when all measures of prophylaxis have been taken and maintained to prevent the recurrence of the disease or its spread to other areas, and when:

In the case of plague, cholera or smallpox, a period of time equal to at least twice the incubation period of the disease, as hereinafter provided, has elapsed since the last case identified has died, recovered or been isolated, and there is no epidemiological evidence of spread of that disease to any contiguous area.

Art. 19 — 2. Every sanitary airport shall have at its disposal:

a) an organised medical service with adequate staff, equipment and premises;

b) facilities for the transport, isolation, and care of infected persons or suspects;

c) facilities for efficient disinfection and disinsecting, for the control of vectors and rodents, and for any other appropriate measure provided for by these Regulations;

d) a bacteriological laboratory, or facilities for dispatching suspected material to such a laboratory;

e) facilities within the airport for vaccination against smallpox, and facilities within the airport or available to it for vaccination against cholera and yellow fever.

Art. 77 — For the purposes of these Regulations the incubation period of smallpox is fourteen days.

Art. 78 — 1. A health administration may require any person on an international voyage who does not show sufficient evidence of protection by a previous attack of smallpox to possess, on arrival, a valid certificate of vaccination against smallpox. Any such person who does not produce such a certificate may be vaccinated or, if he refuses vaccination, he may be placed under surveillance for not more than fourteen days, reckoned from the date of his departure from the last territory visited before arrival.

2. A person on an international voyage, who during a period of fourteen days before his arrival has visited an infected area and who, in the opinion of the health authority, is not sufficiently protected by vaccination or by a previous attack of smallpox, may be required to be vaccinated, or may be placed under surveillance, or may be vaccinated and then placed under surveillance; if he refuses to be vaccinated, he may be isolated. The period of surveillance or isolation shall not be more than fourteen days, reckoned from the date of his departure from the infected area. A valid certificate of vaccination against smallpox shall be considered as evidence of sufficient protection.

3. Any health administration may apply the measures provided for in this Article, whether smallpox infection is present in its territory or not.

Art. 79 — 1. A ship or an aircraft shall be regarded as infected if, on arrival, it has a case of smallpox on board, or if such a case has occurred on board during the voyage.

2. Any other ship or aircraft shall be regarded as healthy even though there may be suspects on board, but any suspect may on disembarking be subjected to the measures provided for in Article 80.

Art. 80 — 1. On arrival of an infected ship or aircraft, the health authority:

a) shall offer vaccination to any person on board who, in its opinion, is not sufficiently protected against smallpox;

b) may, for a period of not more than fourteen days, reckoned from the last exposure to infection, isolate or place under surveillance any person disembarking, but the health authority shall take into account the previous vaccinations of the person and the possibility of his having been exposed to infection in determining the period of such isolation or surveillance;

c) shall disinfect:

i) any baggage of any infected person; and

ii) any other baggage or article such as used bedding or linen, and any part of the ship or aircraft, which is considered to be contaminated.

2. A ship or an aircraft shall continue to be regarded as infected until every infected person has been removed and until the measures required by the health authority in accordance with paragraph 1 of this Article have been effectively carried out. The ship or aircraft shall thereupon be given free pratique.

Art. 81 — On arrival, a healthy ship or aircraft, even when it has come from an infected area, shall be given free pratique.

Art. 82 — If, on arrival of a train, road vehicle or other means of transport, a case of smallpox is discovered, the infected person shall be removed and the provisions of paragraph 1 of Article 80 shall apply, any period of surveillance or isolation being reckoned from the date of arrival, and disinfection being applied to any part of the train, road vehicle or other means of transport which is considered to be contaminated.

CHOLERA

See Smallpox art. 7 and 19

Art. 62 — For the purposes of these Regulations the incubation period of cholera is five days.

Art. 63 — 1. If on arrival of a ship, aircraft, train, road vehicle or other means of transport a case of cholera is discovered, or a case has occurred on board, the health authority *a)* may apply surveillance or isolation of suspects, among passengers or crew for a period not to exceed five days reckoned from the date of disembarkation; *b)* shall be responsible for the supervision of the removal and safe disposal of any water, food (excluding cargo), human dejecta, waste water including bilge water, waste matter, and any other matter which is considered to be contaminated, and shall be responsible for the disinfection of water tanks and food handling equipment.

2. Upon accomplishment of *b)* the ship, aircraft, train, road vehicle or other means of transport shall be given free pratique.

Art. 64 — Foodstuffs carried as cargo on board ships, aircraft, trains, road vehicles or other means of transport in which a case of cholera has occurred during the journey, may not be subjected to bacteriological examination except by the health authorities of the country of final destination.

Art. 65 — 1. No person shall be required to submit to rectal swabbing.

2. A person on an international voyage, who has come from an infected area within the incubation period of cholera and who has symptoms indicative of cholera, may be required to submit to stool examination.